Springer Oceanography

The Springer Oceanography series seeks to publish a broad portfolio of scientific books, aiming at researchers, students, and everyone interested in marine sciences. The series includes peer-reviewed monographs, edited volumes, textbooks, and conference proceedings. It covers the entire area of oceanography including, but not limited to, Coastal Sciences, Biological/Chemical/Geological/Physical Oceanography, Paleoceanography, and related subjects.

More information about this series at http://www.springer.com/series/10175

Andreas Teske · Verena Carvalho
Editors

Marine Hydrocarbon Seeps

Microbiology and Biogeochemistry of a Global Marine Habitat

Springer

Editors
Andreas Teske
Department of Marine Sciences
University of North Carolina
Chapel Hill, NC, USA

Verena Carvalho
Max Planck Institute for Marine Microbiology
Bremen, Germany

ISSN 2365-7677 ISSN 2365-7685 (electronic)
Springer Oceanography
ISBN 978-3-030-34825-0 ISBN 978-3-030-34827-4 (eBook)
https://doi.org/10.1007/978-3-030-34827-4

This Springer imprint is published by the registered company Springer Nature Switzerland AG
The registered company address is: Gewerbestrasse 11, 6330 Cham, Switzerland

Foreword

Hydrocarbon seeps are unlike other seafloor ecosystems. Their dependence on fossil carbon sources, strongly reducing redox state, abundant supply of electron donors (hydrogen, sulfide, methane, ammonia), and frequent brine admixture sets them apart. In this book, aficionados of the oily deep provide an overview on hydrocarbon seeps in major marine regions before directing the spotlight onto some of the special microorganisms thriving in these ecosystems. The book opens with an overview on seep microbial ecology in Chap. 1 by S. Emil Ruff, inspired by his publication "Global dispersion and local diversification of the methane seep microbiome," published in 2015 in Proceedings of the National Academy of Sciences of the United States of America. He and his colleagues established the microbial characteristics of seep ecosystems and delineated their differences to other seafloor microbiota. No other benthic microbial ecosystem is so strongly shaped by methane- and sulfur-cycling bacteria and archaea.

Among the bacteria that thrive in hydrocarbon seeps, perhaps none are as diversified and adaptable as the hydrocarbon-degrading sulfate reducers, introduced, and discussed in depth in Chap. 2 by Sara Kleindienst and Katrin Knittel. Taken together, the wide substrate spectrum and environmental tolerance of this group turn these bacteria into effective and ubiquitous catalysts for hydrocarbon oxidation in anaerobic marine sediments.

In the next five Chaps. (3, 4, 5, 6 and 7), the editors of this book and their colleagues, including Samantha B. Joye, Gunter Wegener, Cassandre S. Lazar, and Konstantinos Ar. Kormas draw inspiration from multiple research cruises and introduce hydrocarbon seeps with contrasting characteristics in different geographical regions to highlight some of the inherent variability that make seep habitats and their microbiota such rewarding targets for microbiological and biogeochemical studies. This gallery of seeps opens with introducing a classic site in the Gulf of California, Guaymas Basin, in Chap. 3—an unusual hybrid location that combines characteristics of a mid-ocean ridge hydrothermal vent site and a hydrocarbon seep. Here, sedimentary organic matter in thick sediments covering an active spreading center is transformed into hydrocarbons under high temperature and pressure. The hydrothermally heated mixture of very young petroleum and hydrothermal energy sources migrates to the sediment surface where it sustains unusually complex microbial communities characterized by extensive thermal tolerance. This chapter also doubles as an introduction to recent advances in alkane utilization and oxidation by novel types of archaea that have been cultured from its hydrothermal sediments, or, that were detected by metagenomic analysis. In contrast to this hydrothermal system, Chap. 4 introduces the most extensive archipelago of cold hydrocarbon seeps on Earth, extending from Florida to Texas along the entire continental slope of the northern Gulf of Mexico. Here, numerous sites represent a wide spectrum of mud volcanoes, seafloor brine lakes, methane hydrate outcrops, and oil seeps that—with few exceptions—remain in the early stages of microbial and biogeochemical surveys and invite further in-depth studies in the future. The widespread hydrocarbon-oxidizing and -assimilating microbial populations in the Gulf of Mexico seeps are of particular environmental significance as they serve as seed banks for hydrocarbon degraders that play a role in natural hydrocarbon remediation, for example, after the Deepwater Horizon oil spill in 2010. A highly unusual seep type in the southern Gulf of Mexico, the Chapopote Asphalt volcano, is singled out in Chap. 5. Instead of gas and liquid hydrocarbon seepage, this site is dominated by the seafloor emergence of slowly flowing, heavily viscous, and chemically weathered asphalt that yet contains sufficient energy sources to sustain its own hydrocarbon seep ecosystem. The habitat surveys are completed by Chaps. 6 and 7 on Mediterranean hydrocarbon seeps, mud volcanoes, and hydrocarbon-rich brine flows, which are embedded in the complex tectonic setting of the Mediterranean Sea, from where they derive some of their distinct qualities.

High concentrations of sulfide, intermixed in the cocktail of reduced gases at seep sites, select for different types of sulfur-oxidizing bacteria, including the highly conspicuous filamentous mat-forming members of the family *Beggiatoaceae*. The filaments are often large enough to be visible to the unaided eye and form extensive white, yellow, and orange-colored mats on the seafloor that coincide with areas of active seepage. At present, these striking bacteria are mostly uncultured; however, the *Beggiatoaceae* are intensively investigated in the context of their habitat. For example, their preferred biogeochemical niche at the sediment–seawater interface is tackled through microprofiler surveys, and physiological experiments are conducted with live mat material. Here, two chapters are devoted to them: In Chap., 8, the two editors of this book summarize and extend current knowledge about the diversity of *Beggiatoaceae* in Gulf of Mexico seeps, while chapter 9, led by Dirk de Beer, opens up a new and previously neglected perspective on the importance of pH and DIC speciation for autotrophic metabolism in *Beggiatoaceae,* using samples collected in Guaymas Basin.

The book concludes with Chap. 10 by Tony Gutierrez and Sara Kleindienst, introducing Stable Isotope Probing (SIP) as a promising and versatile tool to investigate the activities and substrate ranges of hydrocarbon-oxidizing bacteria. SIP elegantly integrates knowledge about the genomic potential of hydrocarbon-oxidizing microorganisms with trophic responses detected in complex mixtures or communities to hydrocarbon availability, using microbial enrichment experiments amended with specific substrates.

Throughout these chapters, all authors have included numerous illustrations and photographs of marine seep ecosystems, keeping in mind that submersible rides and ROV deployments are in short supply, and that many readers have not seen these distinctive seafloor landscapes and microbial habitats in person. Often, images say more than a thousand words; they provide a mental reference and serve as an anchor for the increasingly data-rich studies that are emerging in the productive field of hydrocarbon seep microbiology. An informative and amply illustrated book assists in making the field more accessible because it lowers the "activation energy" barriers that stand in the way of encouraging further interest in deep-sea hydrocarbon seep research. Likewise, this book provides readily accessible background and context for rapidly evolving research "hot spots," for example, the ongoing discoveries of enzymatic pathways of hydrocarbon degradation, the investigation of genomes and metagenomes, the identification of novel hydrocarbon-oxidizing bacteria and archaea, and new insights into stable carbon isotope systematics and molecular structures of metabolites that arise from new findings in the repertoire of microbial degradation of hydrocarbons. For now, this volume provides a useful introduction, anticipating to stimulate further research interest in hydrocarbon seep microbiology and biogeochemistry.

Chapel Hill, USA — Andreas Teske
Bremen, Germany — Verena Carvalho

Acknowledgements

Andreas Teske is thanking the Hanse Institute for Advanced Studies (Hanse-Wissenschaftskolleg Delmenhorst) for providing a perfect environment that is conducive to making plans and writing, while remaining in close reach and communication to many authors of chapters in this book. His research on hydrocarbon seep microbiology is supported by the National Science Foundation (Programs: Microbial Interactions and Processes/Microbial Observatories, Biological Oceanography, Molecular and Cellular Biosciences). Verena Carvalho was supported by the German Research Foundation fellowship SA2505/1-1, and the Max Planck Society.

Contents

Chapter 1
Microbial Communities and Metabolisms at Hydrocarbon Seeps

S. Emil Ruff

Abstract Hydrocarbon seeps are common features of all oceans and are located mainly along the continental margins (Fig. 1). Seeps are locally restricted, yet highly productive hotspots of biodiversity that experience very different environmental conditions and energy regimes than the surrounding deep-sea sediments. Hydrocarbon seep ecosystems are mostly fueled by methane. Occasionally, seeps are found that emit the short-chain hydrocarbons ethane, propane or butane, and even oil and asphalt seeps have been described. Seep ecosystems therefore comprise ecological niches and microbial clades that are distinct from those found in deep-sea sediments, which are not fuelled by methane and other hydrocarbons. This chapter provides an overview of the communities thriving at marine hydrocarbon seeps and the microbial metabolisms that create these oases of life (with references to other chapters in this book). It highlights the current knowledge of the diversity and biogeography of seep microbial communities and presents possible mechanisms governing their community assembly.

Life at Hydrocarbon Seeps

Due to the abundant energy supply from subsurface sources of hydrocarbons, cold seeps sustain microbial and faunal communities with population sizes that can exceed those found in the surrounding marine sediments by several orders of magnitude. A teaspoon of seep sediment may contain tens of billions of microbial cells (10^{10} cells ml^{-1}; Ruff et al. 2013; Marlow et al. 2014), and a square meter of surface sediment can be covered by tens of thousands of individual animals, such as ampharetid polychaetes (Sommer et al. 2010). The microbial communities of seep sediments belong to the biomass-richest in nature, with cell numbers comparable to those found in soil (Portillo et al. 2013) and the human gut (Sender et al. 2016). This suggests an efficient conversion of available energy sources, but also the existence of

S. Emil Ruff (✉)
Marine Biological Laboratory, Woods Hole, MA, USA
e-mail: eruff@mbl.edu

A. Teske and V. Carvalho (eds.), *Marine Hydrocarbon Seeps*,
Springer Oceanography, https://doi.org/10.1007/978-3-030-34827-4_1

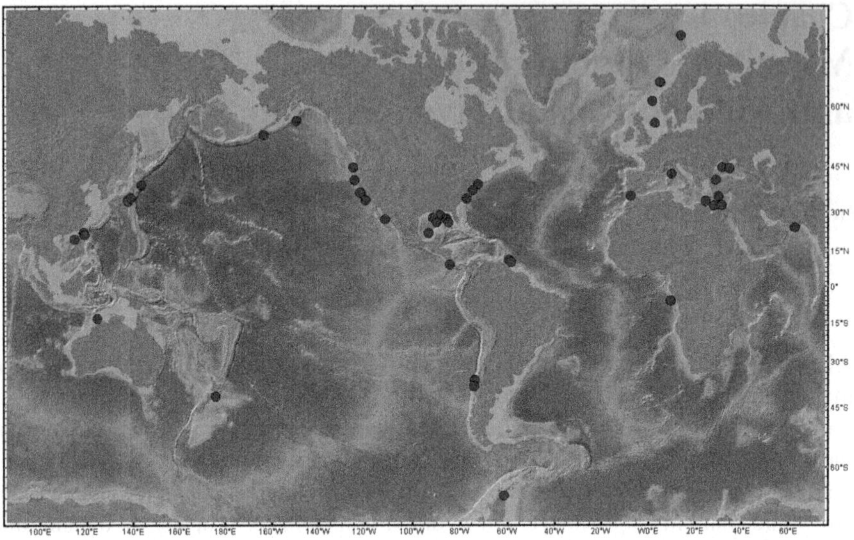

Fig. 1 Map of seafloor regions with investigated marine hydrocarbon seepage (orange dots). The map shows a selection of well-known seep areas. The total number of seeps worldwide is unknown, but estimated to be at least several ten thousand. The map was created using GeoMapApp

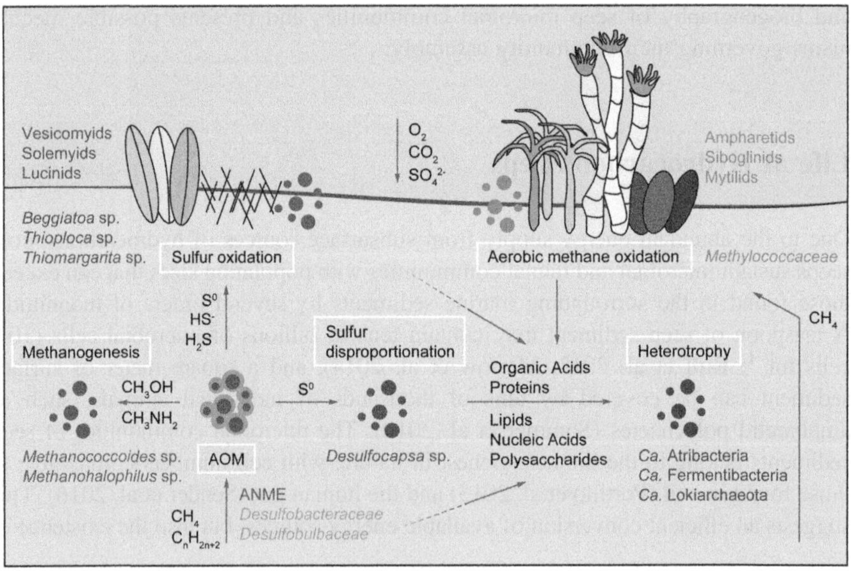

Fig. 2 Schematic overview of the food web and the microbial metabolisms found at hydrocarbon seeps. Methane and hydrocarbon oxidizers are at the base of the food web turning subsurface-derived energy into biomass. Their metabolic activity creates ecological niches for other microorganisms by producing methylated compounds used by methanogens, and sulfur compounds used by sulfur oxidizers and sulfur disproportionators. The organic matter build up by methane- or sulfur-cycling microbes is then used by heterotrophs, further increasing the diversity and complexity of the ecosystem. Note: The species that are associated with a metabolism serve as examples, e.g. *Beggiatoa sp.* are common sulfur oxidizers, and do not represent a comprehensive list

a complex food web with numerous community functions (Fig. 2). At the center of this food web are chemosynthetic microorganisms. These organisms are the primary producers of biomass and catalyze vital ecosystem functions such as aerobic and anaerobic methanotrophy, sulfate reduction, and sulfide oxidation (see, for example, organisms presented in Chaps. 2, 5, and 9). Chemosynthetic microorganisms harness the energy of reduced chemical compounds like hydrogen, sulfide and methane. The microbes use this energy to fix inorganic carbon and synthesize organic carbon compounds and cellular building blocks such as proteins, sugars, lipids, and nucleic acids. The organic carbon produced by chemosynthetic populations is then used by heterotrophic microorganisms or animals in the food web (Fig. 2). Chemosynthesis is performed by free-living microorganisms in the sediment or by host-associated microorganisms living in symbiosis with animals, including mussels, clams, and tube worms (Figs. 2 and 3). The most well-known animal hosts at seeps are siboglinid tubeworms, bathymodiolin mussels, and vesicomyid clams (Fig. 3), which live in

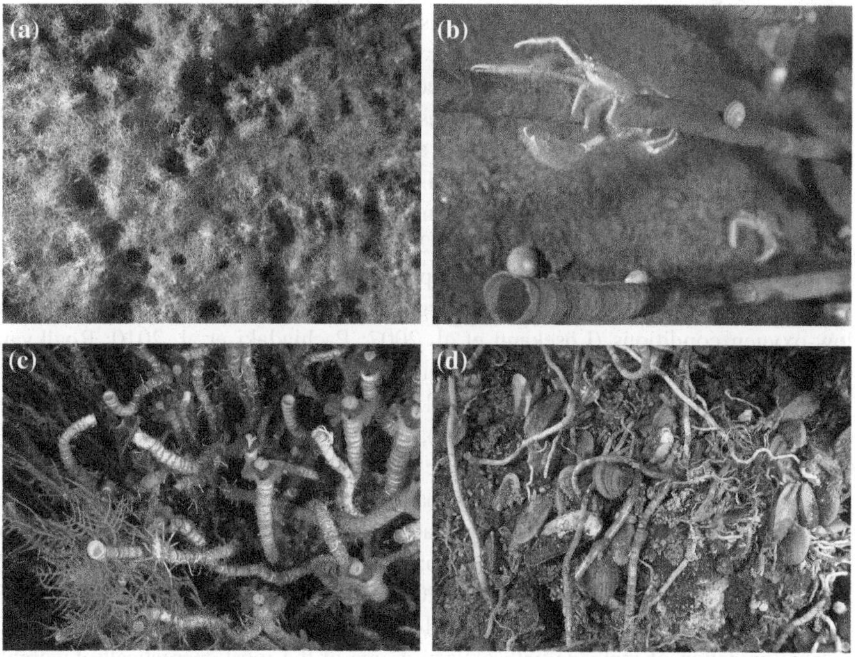

Fig. 3 Seep-associated microbiota and macrofauna. **a** Orange mats are composed of giant filamentous bacteria that oxidize sulfide at the sediment-water interface; **b** Deep-sea crabs are often covered with white filaments of sulfur-oxidizing bacteria growing on their carapace; **c** *Lamellibrachia* tubeworms within the family *Siboglinidae* grow in dense patches and can live for centuries. They lack a mouth and gut, and are dependent on nutrition provided by sulfur-oxidizing endosymbionts. **d** Mytilid mussels of the genus *Bathymodiolus* live on the sediment surface and harbor sulfur- and methane-oxidizing endosymbionts. (Images courtesy of I.R. MacDonald, Florida State University)

symbiosis with methane-oxidizing or sulfur-oxidizing bacteria (Cordes et al. 2005; Levin 2005; Decker et al. 2013).

Life Styles at Hydrocarbon Seeps

Aerobic Methanotrophy

Aerobic methanotrophic bacteria use oxygen to consume methane and are found in the upper, mostly oxic sediment layers of hydrocarbon seeps (Yan et al. 2006; Lösekann et al. 2007; Tavormina et al. 2008; Wasmund et al. 2009). At seeps with a high fluid flow this oxic layer is only a few millimeters thick. Sediment surfaces may become completely anoxic when the seafloor is covered by microbial mats of sulfur-oxidizing bacteria that consume the oxygen before it penetrates into the sediment (Fig. 3a). More details about mat-forming sulfide-oxidizing bacteria are presented in Chaps. 8 and 9. With decreasing fluid flow and increasing bioirrigation by macrofauna, the thickness of the oxic layer can increase to several centimeters. Aerobic methanotrophs often occur in sediments that are not covered by microbial mats, and are bioirrigated by tubeworms (Sommer et al. 2010; Ruff et al. 2013; Thurber et al. 2013), or in sediments that are frequently disturbed, such as the active centers of mud volcanoes (Niemann et al. 2006; Felden et al. 2010, 2013). Aerobic methanotrophs are generally less abundant in seeps with low and stable fluid fluxes, e.g., above methane hydrates (Ruff et al. 2015). However, the occurrence of aerobic methanotrophic *Methylococcales* in hypoxic and even anoxic/sulfidic layers at hydrocarbon seeps indicates that these organisms can also adapt to high-sulfide and low-oxygen conditions (Lösekann et al. 2007; Pachiadaki et al. 2010; Roalkvam et al. 2011; Ruff et al. 2013). Some aerobic methanotrophs are even active in anoxic waters of the Black Sea (Blumenberg et al. 2007), present in oxygen minimum zones of the Pacific Ocean (Hayashi et al. 2007), and in anoxic peat soils (Roslev and King 1994), surviving under prolonged anoxic conditions (Roslev and King 1995).

Seep ecosystems seem to harbor mainly type-I methanotrophs of the order *Methylococcales*, which cluster among at least three distinct phylogenetic groups (Fig. 4 red clades) (Inagaki et al. 2004b; Yan et al. 2006; Lösekann et al. 2007; Tavormina et al. 2008; Wasmund et al. 2009; Foucher et al. 2010; Ruff et al. 2013, 2015; Thurber et al. 2013; Felden et al. 2014; Oshkin et al. 2014; Paul et al. 2017). The aerobic oxidation of methane with oxygen is one of the most energetically favorable reactions in the microbial realm, yielding up to 800 kJ per mol methane oxidized. The type-I methanotrophs use the ribulose monophosphate (RuMP) pathway, a highly efficient metabolism for fixing carbon (Kato et al. 2006), which likely provides an advantage at seep systems. Hence, these aerobic methanotrophs are relatively fast growing, and are among the early colonizers of seep ecosystems (Ruff et al. 2019). The use of alternative electron acceptors, such as nitrate, or novel pathways to perform methane oxidation in anoxic habitats, has been indicated in the order *Methylococcales* (Costa et al. 2017; Martinez-Cruz et al. 2017), which may explain the occurrence of

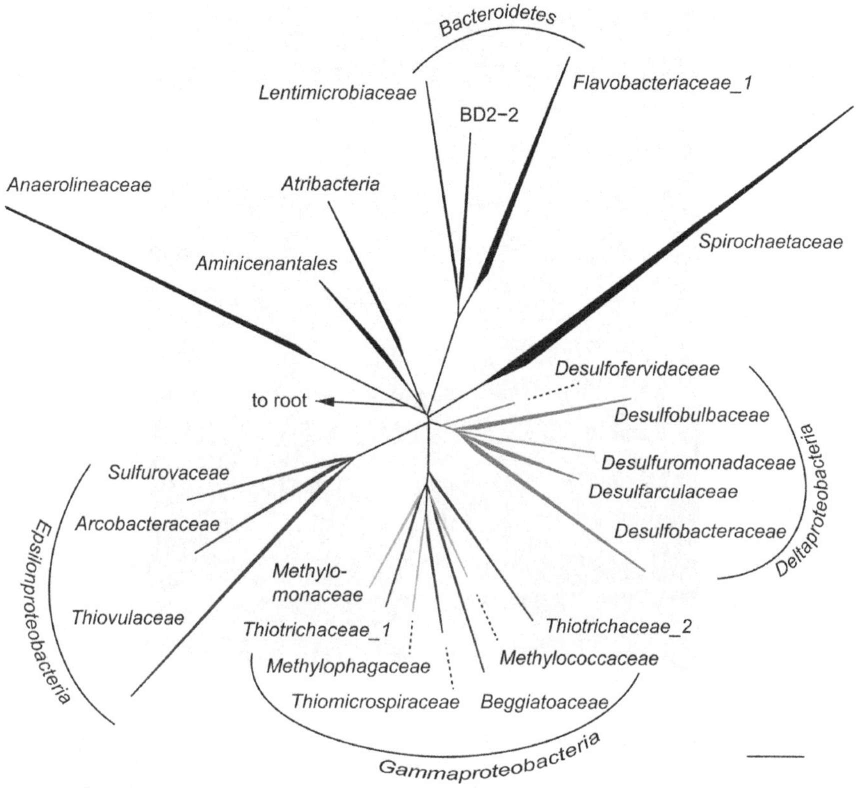

Fig. 4 16S rRNA gene based phylogenetic tree showing common seep-associated bacteria, color-coded according to their energy metabolism: red = aerobic methane oxidizers, blue = sulfur oxidizers, brown = sulfate reducers, black = heterotrophs. Scale bar: 10% estimated sequence divergence

active aerobic methanotrophs in anoxic methane seep sediments (Ruff et al. 2013). At seep ecosystems, aerobic methanotrophs are also common as symbionts of mussels (Duperron et al. 2008, 2011) and siboglinids (Dubilier et al. 2008; Hilário et al. 2011) (Fig. 2). The animals shelter the microorganisms in specialized organs or tissues, providing them with oxygen and methane. In return, the microorganisms provide reduced organic carbon compounds for their hosts.

Anaerobic Methanotrophy Coupled to Sulfate Reduction

Anaerobic methanotrophic archaea (ANME) dominate anoxic sulfidic sediments and typically occur in consortia with sulfate-reducing bacteria (SRB) (Fig. 5). These consortia perform the anaerobic oxidation of methane (AOM) coupled with

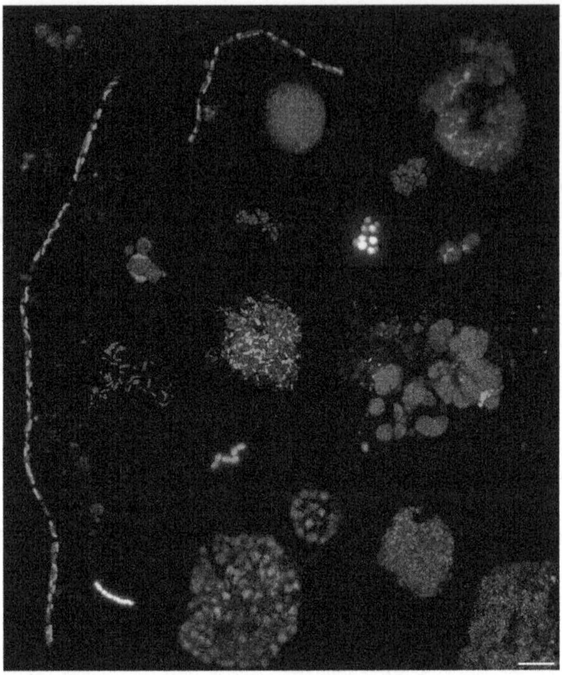

Fig. 5 Compilation of fluorescence micrographs of anaerobic methanotrophic archaea (red = ANME) and sulfate-reducing bacteria (green = SRB), which play key roles in the anaerobic oxidation of methane in marine sediments. The organisms can occur by themselves, but are mostly found together, forming syntrophic consortia. Aerobic methanotrophic *Methylococcales* (beige) do not need a partner to metabolize methane. Close relatives of free-living *Methylococcales* are found as chemosynthetic symbionts of marine invertebrates (Petersen and Dubilier 2009). Together, aerobic and anaerobic methanotrophs remove around 30% of the methane emitted from the seafloor of the world's oceans, and hence greatly impact the global methane budget (Boetius and Wenzhöfer 2013). Scale bar = 20 μm. *Figure credit* Katrin Knittel, S. Emil Ruff

sulfate reduction (SR) (Knittel and Boetius 2009). AOM in Guaymas Basin sediments is presented in Chap. 3.

The AOM net reaction

$$CH_4 + SO_4^{2-} \rightarrow HCO_3^- + HS^- + H_2O$$

yields between −20 and −40 kJ per mol methane in situ, and AOM is thus one of the least favorable energy-conserving catabolic reactions in nature. The produced carbonate often precipitates on the seafloor, forming chemoherms that are used as substrate for macrofauna. The produced sulfide is an energy-rich compound that diffuses upwards creating niches for sulfide-oxidizing bacteria at the sediment surface (for more details refer to Chaps. 8 and 9).

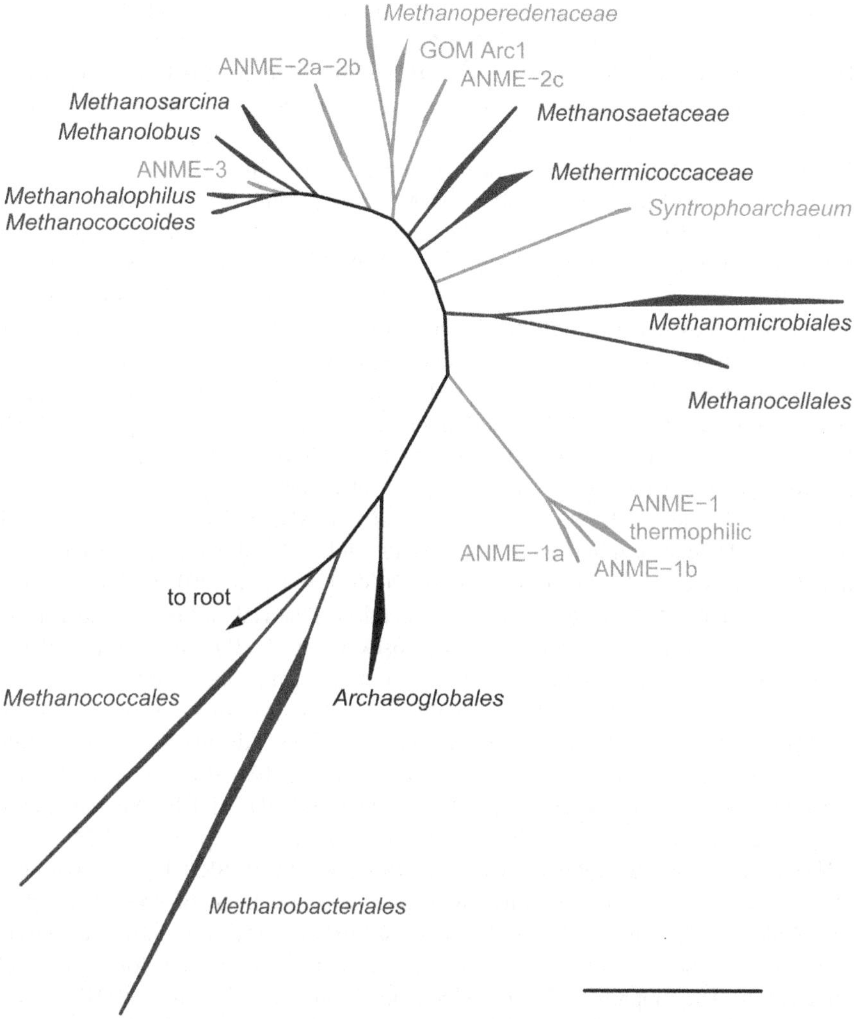

Fig. 6 16S rRNA gene based phylogeny of ANME archaea showing the affiliation of anaerobic hydrocarbon-oxidizing archaea (red) in relation to their methanogenic relatives (blue) within the class *Methanomicrobia*. Scale bar: 10% estimated sequence divergence

ANME belong to the class *Methanomicrobia* and affiliate with three major clades: ANME-1, ANME-2 and ANME-3 (Hinrichs et al. 1999; Orphan et al. 2002; Niemann et al. 2006) (Fig. 6). These major clades comprise the subgroups ANME-1a, ANME-1b, thermophilic ANME-1, as well as ANME-2a-d (Orphan et al. 2001; Teske et al. 2002; Meyerdierks et al. 2010; Holler et al. 2011; Biddle et al. 2012; Merkel et al. 2012; Haroon et al. 2013). The phylogenetically broadest clade is ANME-1, forming an order now named *Methanophagales* within the class *Methanomicrobia* (Adam et al. 2017). ANME-1 predominate in deeper sediment layers that are often characterized by high sulfide and low sulfate concentrations (Elvert et al. 2005; Knittel et al. 2005; Niemann et al. 2005; Wegener et al. 2008; Roalkvam et al. 2011; Yanagawa et al. 2011; Vigneron et al. 2013). A thermophilic subgroup of ANME-1 occurs in hydrothermal sediments (Teske et al. 2002; Holler et al. 2011; Biddle et al. 2012) and hydrothermal fluids (Merkel et al. 2012). Members of ANME-1 have also been found in hypersaline seep sediments of the Gulf of Mexico (Lloyd et al. 2006). ANME-2 is a family-level group within the *Methanosarcinales*. They occur preferentially in sulfate-penetrated, sulfide- and methane-rich surface sediments (Elvert et al. 2005; Knittel et al. 2005; Wegener et al. 2008; Rossel et al. 2011; Yanagawa et al. 2011). ANME-2a are the most common and globally most widespread ANME clade (Ruff et al. 2015). They seem to have an advantage at sites where sulfate and methane concentrations are high, e.g., above shallow hydrates (Knittel et al. 2005; Lösekann et al. 2007) and in enrichment cultures (Wegener et al. 2016), and likely outcompete other mesophilic ANME lineages due to their faster growth rate. ANME-2c prefer low methane fluxes and bioturbated sediments, e.g., at seeps that are inhabited by chemosynthetic mussels and clams (Knittel et al. 2005; Felden et al. 2014). ANME-3 is a genus-level clade within the *Methanosarcinales*, and despite its global distribution, this group rarely dominates in seep sediments (Knittel et al. 2005; Lazar et al. 2011; Vigneron et al. 2013; Ruff et al. 2015). To date, the only exception is the Håkon Mosby mud volcano in the Barents Sea (Niemann et al. 2006; Lösekann et al. 2007; Ruff et al. 2019).

Sulfate reducers are commonly found in marine sediments, as sulfate is a ubiquitous electron acceptor used in the degradation of hydrocarbons or organic matter (Ravenschlag et al. 2001; Muyzer and Stams 2008). The SRB involved in mesophilic AOM belong to the class *Deltaproteobacteria* and are subdivided into several clades related to the genera *Desulfosarcina* (DSS) or *Desulfobulbus* (DBB) (Schreiber et al. 2010; Kleindienst et al. 2012) (Fig. 7). At the majority of seep sites, ANME-1 and ANME-2 are found to aggregate with the DSS clade SEEP-SRB-1 (Schreiber et al. 2010), whereas ANME-3 mainly aggregate with relatives of the DBB clade SEEP-SRB-3 (Niemann et al. 2006). These associations, however, do not seem to be exclusive, and the diversity of aggregates and mechanisms of aggregation are not well understood. Thermophilic ANME, which are a subgroup within ANME-1, form consortia with sulfate reducers affiliating with the clade HotSeep-1 (Wegener et al. 2015). The first organism of this clade "*Candidatus* Desulfofervidus auxilii" was recently cultured and described (Krukenberg et al. 2016).

Anaerobic Methanotrophy Coupled to Nitrate, Nitrite, and Metal Oxides

The anaerobic oxidation of methane coupled with sulfate reduction was a long-standing geochemical enigma until it was shown that syntrophic consortia of anaerobic methanotrophic archaea and sulfate-reducing bacteria mediate this process (Hinrichs et al. 1999; Boetius et al. 2000). In recent years, however, evidence has emerged that methane can also be oxidized anaerobically using alternative electron acceptors. The newly described family "*Candidatus* Methanoperedenaceae" (related to ANME-2d) contains organisms that couple the anaerobic oxidation of methane to the reduction of nitrate (Haroon et al. 2013). "*Candidatus* Methylomirabilis oxyfera" belong to the bacterial candidate phylum NC10, and couple methane oxidation to the reduction of nitrite (Ettwig et al. 2010). Both processes are considered of minor importance at marine seeps since nitrate and nitrite are less abundant than the ubiquitous marine electron acceptor sulfate. However, these metabolisms appear to be widespread in freshwater ecosystems (Hu

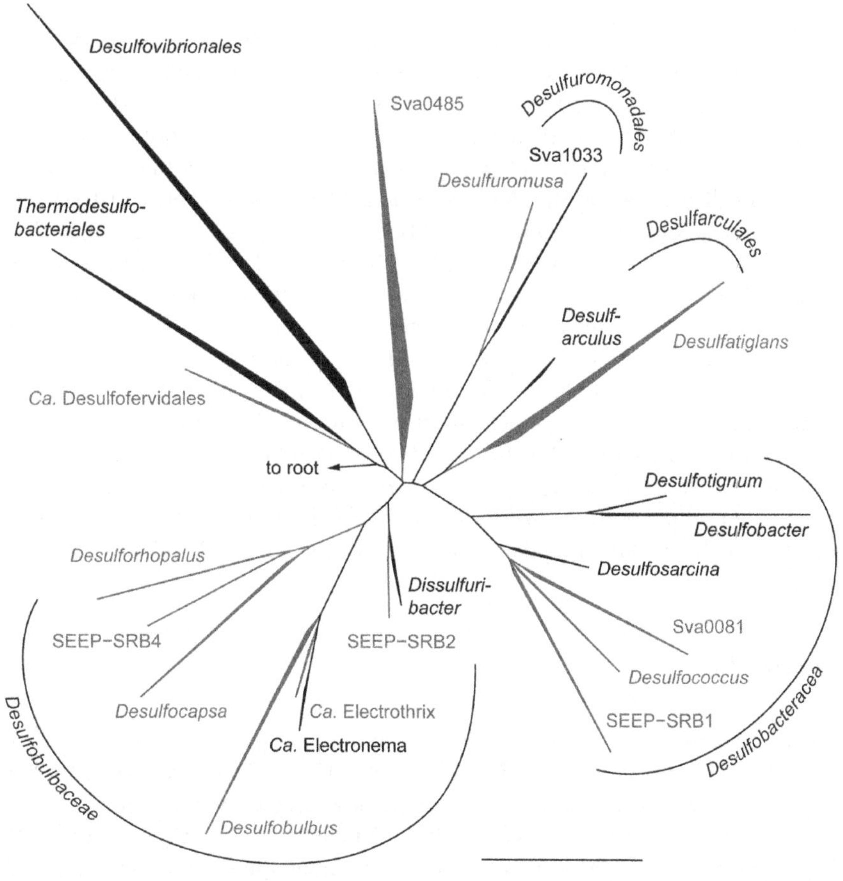

Fig. 7 16S rRNA gene based phylogeny of seep-associated *Deltaproteobacteria* (brown). *Desulfosarcina* relatives of the SEEP-SRB-1 cluster and *Desulfobulbus* relatives have been shown to associate with ANME-1, ANME-2, and ANME-3. Seep sediments also harbor *Deltaproteobacteria* that are not found in consortia with ANMEs and are common in other marine ecosystems, for example *Ca.* Electrothrix and the uncultured clade Sva0485. Scale bar: 10% estimated sequence divergence

et al. 2014; Vigneron et al. 2017; Weber et al. 2017), and in permafrost ecosystems (Winkel et al. 2018). The clade ANME-2d also includes organisms that can couple methane oxidation to the reduction of manganese- and iron-oxides (Ettwig et al. 2016; Oni and Friedrich 2017), a process that has been found to occur in coastal sediments (Egger et al. 2015) as well as in hydrocarbon seeps and hydrothermal sediments (Beal et al. 2009; Wankel et al. 2012; Sivan et al. 2014).

Aerobic and Anaerobic Oxidation of Non-methane Hydrocarbons

Although methane is generally the most abundant hydrocarbon at seeps, higher hydrocarbons can be used by free-living or symbiotic microorganisms in particular at thermogenic hydrocarbon seeps. At these seeps, the measured rates of sulfate reduction exceed those of methane oxidation (Orcutt et al. 2005), and diverse assemblages of *Archaea* and *Deltaproteobacteria* have been identified that degrade short-chain alkanes aerobically (Mastalerz et al. 2009; Redmond et al. 2010; Li et al. 2013) and anaerobically (Adams et al. 2013; Bose et al. 2013; Jaekel et al. 2013; Kleindienst et al. 2014; Dowell et al. 2016; Stagars et al. 2016). Recently, several major findings have improved our understanding of these communities. Laso-Pérez and colleagues described a consortium of the archaeon "*Candidatus* Syntrophoarchaeum" (formerly Gom-Arc87) and the partner bacterium "*Candidatus* Desulfofervidus" that couples the anaerobic degradation of short-chain alkanes with sulfate reduction, relying on biochemical mechanisms described for methanogens (Laso-Pérez et al. 2016). Rubin-Blum and colleagues showed for the first time that short-chain alkane-degrading *Cycloclasticus* are symbionts of seep-associated mussels and sponges (Rubin-Blum et al. 2017). Further details about microbial non-methane hydrocarbon oxidation are presented in Chaps. 2 and 5.

Thiotrophy

Seep-associated sulfide oxidizers—or thiotrophs—belong mostly to the gammaproteobacterial family *Beggiatoaceae* (Fig. 4 blue clades), and form extensive mats of white, yellow, or orange filaments on top of seep sediments (Fig. 3a; Joye et al. 2004; Mills et al. 2004; Knittel et al. 2005; Lloyd et al. 2010; Grünke et al. 2012; Meyer et al. 2013). The giant filamentous bacteria of the *Beggiatoaceae* are conspicuous and visible to the naked eye, as they can cover square meters of sediment. They use oxygen or nitrate to oxidize sulfide that is produced during AOM in deeper sediment layers (Preisler et al. 2007). Non-filamentous *Beggiatoaceae*, such as *Thiomargarita* and "*Candidatus* Thiopilula", also occur at seeps, but are observed less frequently (Girnth et al. 2011;

Jones et al. 2015). Some seeps harbor mats of sulfur-oxidizing bacteria within the epsilonproteobacterial order *Campylobacterales*, such as *Arcobacter* (Omoregie et al. 2008; Grünke et al. 2011) and *Sulfurovum* (Inagaki et al. 2004a; Roalkvam et al. 2011). The predominance of different sulfur oxidizers has been attributed to temporal and spatial variations in energy supply (Grünke et al. 2011). Members of the *Campylobacterales*, for instance, have been found at many seeps (Campbell et al. 2006), yet, they seem to dominate only a few habitats, e.g., a very deep seep ecosystem in the Japan Trench that is covered with syntrophic vesicomyid clams (Felden et al. 2014). Whether this is a result of niche preferences of the clades or due to competitive advantages favoring thiotrophic *Gammaproteobacteria* remains to be elucidated. Sulfide-oxidizing bacteria are also common symbionts of marine invertebrates such as mussels, clams, and siboglinids at hydrocarbon seeps (Dubilier et al. 2008). Further details about mat-forming sulfide-oxidizing bacteria are presented in Chaps. 8 and 9.

Heterotrophy

Hydrocarbon seeps also harbor a plethora of archaeal and bacterial clades that have a heterotrophic lifestyle, using organic carbon compounds as energy sources. Among these organisms, some are found in greater abundance at seeps than in the surrounding deep-sea sediments, and hence may belong to a specific seep-associated microbiome (Fig. 4, black clades; Ruff et al. 2015), yet their ecological niches remain largely elusive. Among the heterotrophic *Archaea*, seep sediments are colonized by members of the *Bathyarchaeota* (formerly Miscellaneous Crenarchaeotic Group), *Lokiarchaeota* (formerly Marine Benthic Group B), *Thermoprofundales* (formerly Marine Benthic Group D) and *Thermoplasmatales* (Vigneron et al. 2013, 2014; Ruff et al. 2015; Yoshinaga et al. 2015; Cui et al. 2016; Cho et al. 2017; Zhou et al. 2019). Heterotrophic seep-associated *Bacteria* include members of the candidate phyla "Atribacteria" (formerly JS1/OP9), and "Fermentibacteria" (formerly Hyd24-12), the bacterial phyla *Planctomycetes*, *Spirochaeta*, *Deferribacteres*, *Chloroflexi*, and the classes *Alphaproteobacteria* and *Betaproteobacteria* (Mills et al. 2005; Pernthaler et al. 2008; Harrison et al. 2009; Knittel and Boetius 2009; Chevalier et al. 2013; Vigneron et al. 2013; Ruff et al. 2015; Trembath-Reichert et al. 2016). Members of the *Spirochaeta* occur as enigmatic endosymbionts of marine oligochaetes (Blazejak et al. 2005), or perform sulfide oxidation in consortial association with sulfate-reducing bacteria (Dubinina et al. 2011). The phylum *Deferribacteres* includes iron-, sulfur-, and nitrate-reducing bacteria (Janssen et al. 2002; Gittel et al. 2012). Members of the candidate division "Fermentibacteria" were originally found at a Hydrate Ridge methane seep (Knittel et al. 2003) but occur at most cold seeps as part of the seep core microbiome (Ruff et al. 2015). "*Ca.* Fermentibacteria" are likely involved in fermentation of decaying biomass (Kirkegaard et al. 2016; Saad et al. 2017). "*Ca.* Atribacteria" are also frequently found at seep ecosystems and it has been proposed that they are involved in methane cycling at seep ecosystems (Zhang et al. 2012; Chevalier et al. 2013). The role of heterotrophic microorganisms and microbial eukaryotes in carbon cycling and syntrophy, however, needs to be further investigated, especially since these organisms account for most of the biodiversity in these ecosystems.

Microbial Community Assembly and Biogeography

Cold seeps are globally distributed, highly productive oases of life, and substantially contribute to seafloor biodiversity. Yet, individual seeps usually cover small areas of a few square meters of seafloor and are widely scattered across large distances. The seeps are thus very localized ecosystems, disconnected by vast areas of bare seafloor. Nevertheless, the seeps of the world's oceans share a large number of microorganisms that are phylogenetically similar and functionally identical (Ruff et al. 2015). The mechanisms driving the global distribution of the seep microbiome remain to be fully characterized. Free-living aerobic methane- and sulfur oxidizers, as well as aerobic heterotrophs, likely disperse with bottom water currents, while host-associated aerobes may hitchhike between the seeps with their hosts. Anaerobic organisms, such as anaerobic methanotrophs and sulfate reducers, however, are often oxygen sensitive; dispersal of these organisms through oceanic currents is therefore less likely. Identification of the mechanisms of their distribution is, as a result, much more elusive.

First insights into the timescales and mechanisms of community assembly suggest that aerobic methanotrophs—having high energy yields and high growth rates—are early colonizers of young seep ecosystems, and of freshly erupted sediments from mud volcanoes (Sommer et al. 2009; Felden et al. 2013; Ruff et al. 2019). AOM consortia—having much lower energy yields and slower growth rates—possess the advantage of thriving in deeper, anoxic sediment layers, thereby consuming methane before it can reach the sediment surface. In developing seep systems, anaerobic methanotrophs appear to slowly outcompete aerobic methanotrophs from below, one reason why *Methylococcales* are often limited to a very thin layer at the surface. The sulfide that is produced by AOM is used as an energy source by thiotrophs, thus creating new ecological niches on the seafloor. The large amount of biomass produced by the autotrophic organisms in turn provides organic carbon and additional niches for heterotrophic organisms. It can take many years to decades before a complex seep microbiome with all its ecological niches, biodiversity, and community functions is established (Fig. 2; Ruff et al. 2019). In particular, the roles of heterotrophs, microbial eukaryotes, and viruses in these systems are not well understood and merit further research.

References

Adam PS, Borrel G, Brochier-Armanet C, Gribaldo S (2017) The growing tree of Archaea: new perspectives on their diversity, evolution and ecology. ISME J 11:2407–2425

Adams M, Hoarfrost A, Bose A, Joye S, Girguis P (2013) Anaerobic oxidation of short-chain alkanes in hydrothermal sediments: potential influences on sulfur cycling and microbial diversity. Front Microbiol 4:110

Beal EJ, House CH, Orphan VJ (2009) Manganese- and iron-dependent marine methane oxidation. Science 325:184–187

Biddle JF, Cardman Z, Mendlovitz H, Albert DB, Lloyd KG, Boetius A, Teske A (2012) Anaerobic oxidation of methane at different temperature regimes in Guaymas Basin hydrothermal sediments. ISME J 6:1018–1031

Blazejak A, Erséus C, Amann R, Dubilier N (2005) Coexistence of bacterial sulfide oxidizers, sulfate reducers, and spirochetes in a gutless worm (Oligochaeta) from the Peru Margin. Appl Environ Microbiol 71:1553–1561

Blumenberg M, Seifert R, Michaelis W (2007) Aerobic methanotrophy in the oxic-anoxic transition zone of the Black Sea water column. Org Geochem 38:84–91

Boetius A, Ravenschlag K, Schubert CJ, Rickert D, Widdel F, Gieseke A et al (2000) A marine microbial consortium apparently mediating anaerobic oxidation of methane. Nature 407:623–626

Boetius A, Wenzhöfer F (2013) Seafloor oxygen consumption fuelled by methane from cold seeps. Nat Geosci 6:725–734

Bose A, Rogers D, Adams M, Joye S, Girguis P (2013) Geomicrobiological linkages between short-chain alkane consumption and sulfate reduction rates in seep sediments. Front Microbiol 4:386

Campbell B, Engel AS, Porter ML, Takai K (2006) The versatile epsilon-proteobacteria: key players in sulphidic habitats. Nat Rev Microbiol 4:458–468

Chevalier N, Bouloubassi I, Birgel D, Taphanel MH, López-García P (2013) Microbial methane turnover at Marmara Sea cold seeps: a combined 16S rRNA and lipid biomarker investigation. Geobiology 11:55–71

Cho H, Hyun J-H, You O-R, Kim M, Kim S-H, Choi D-L et al (2017) Microbial community structure associated with biogeochemical processes in the sulfate–methane transition zone (SMTZ) of gas-hydrate-bearing sediment of the Ulleung Basin, East Sea. Geomicrobiol J 34:207–219

Cordes EE, Arthur MA, Shea K, Arvidson RS, Fisher CR (2005) Modeling the mutualistic interactions between tubeworms and microbial consortia. PLoS Biol 3:e77

Costa RB, Okada DY, Martins TH, Foresti E (2017) Aerobic methanotrophs grew under anoxic conditions and supported a diverse heterotrophic bacterial community. Environ Eng Sci 35:804–814

Cui H, Su X, Chen F, Wei S, Chen S, Wang J (2016) Vertical distribution of archaeal communities in cold seep sediments from the Jiulong methane reef area in the South China Sea. Biosci J 32:4

Decker C, Olu K, Cunha RL, Arnaud-Haond S (2013) Phylogeny and diversification patterns among vesicomyid bivalves. PLoS ONE 7:e33359

Dowell F, Cardman Z, Dasarathy S, Kellermann M, Lipp JS, Ruff SE, Biddle JF, McKay LJ, MacGregor BJ, Lloyd KG, Albert DB, Mendlovitz H, Hinrichs KU, Teske A (2016) Microbial communities in methane- and short chain alkane-rich hydrothermal sediments of Guaymas Basin. Front Microbiol 7:17

Dubilier N, Bergin C, Lott C (2008) Symbiotic diversity in marine animals: the art of harnessing chemosynthesis. Nat Rev Microbiol 6:725–740

Dubinina G, Grabovich M, Leshcheva N, Rainey FA, Gavrish E (2011) *Spirochaeta perfilievii* sp. nov., an oxygen-tolerant, sulfide-oxidizing, sulfur- and thiosulfate-reducing spirochaete isolated from a saline spring. Int J Syst Evol Microbiol 61:110–117

Duperron S, Halary S, Lorion J, Sibuet M, Gaill F (2008) Unexpected co-occurrence of six bacterial symbionts in the gills of the cold seep mussel Idas sp. (Bivalvia: *Mytilidae*). Environ Microbiol 10:433–445

Duperron S, Guezi H, Gaudron SM, Pop Ristova P, Wenzhöfer F, Boetius A (2011) Relative abundances of methane- and sulphur-oxidising symbionts in the gills of a cold seep mussel and link to their potential energy sources. Geobiology 9:481–491

Egger M, Rasigraf O, Sapart CJ, Jilbert T, Jetten MSM, Röckmann T et al (2015) Iron-mediated anaerobic oxidation of methane in brackish coastal sediments. Environ Sci Technol 49:277–283

Elvert M, Hopmans EC, Treude T, Boetius A, Suess E (2005) Spatial variations of methanotrophic consortia at cold methane seeps: implications from a high-resolution molecular and isotopic approach. Geobiology 3:195–209

Ettwig KF, Butler MK, Le Paslier D, Pelletier E, Mangenot S, Kuypers MMM et al (2010) Nitrite-driven anaerobic methane oxidation by oxygenic bacteria. Nature 464:543–548

Ettwig KF, Zhu B, Speth D, Keltjens JT, Jetten MSM, Kartal B (2016) Archaea catalyze iron-dependent anaerobic oxidation of methane. Proc Natl Acad Sci U S A 113:12792–12796

Felden J, Lichtschlag A, Wenzhöfer F, de Beer D, Feseker T, Pop Ristova P et al (2013) Limitations of microbial hydrocarbon degradation at the Amon mud volcano (Nile deep-sea fan). Biogeosciences 10:3269–3283

Felden J, Ruff SE, Ertefai T, Inagaki F, Hinrichs K-U, Wenzhöfer F (2014) Anaerobic methanotrophic community of a 5346-m-deep vesicomyid clam colony in the Japan Trench. Geobiology 12:183–199

Felden J, Wenzhöfer F, Feseker T, Boetius A (2010) Transport and consumption of oxygen and methane in different habitats of the Håkon Mosby Mud Volcano (HMMV). Limnol Oceanogr 55:2366–2380

Foucher J-P, Dupré S, Scalabrin C, Feseker T, Harmegnies F, Nouzé H (2010) Changes in seabed morphology, mud temperature and free gas venting at the Håkon Mosby mud volcano, offshore northern Norway, over the time period 2003–2006. Geo-Marine Lett 30:157–167

Girnth AC, Grünke S, Lichtschlag A, Felden J, Knittel K, Wenzhofer F et al (2011) A novel, mat-forming *Thiomargarita* population associated with a sulfidic fluid flow from a deep-sea mud volcano. Environ Microbiol 13:495–505

Gittel A, Kofoed MVW, Sørensen KB, Ingvorsen K, Schramm A (2012) Succession of Deferribacteres and Epsilonproteobacteria through a nitrate-treated high-temperature oil production facility. Syst Appl Microbiol 35:165–174

Grünke S, Felden J, Lichtschlag A, Girnth A-C, De Beer D, Wenzhöfer F, Boetius A (2011) Niche differentiation among mat-forming, sulfide-oxidizing bacteria at cold seeps of the Nile Deep Sea Fan (Eastern Mediterranean Sea). Geobiology 9:330–348

Grünke S, Lichtschlag A, de Beer D, Felden J, Salman V, Ramette A et al (2012) Mats of psychrophilic thiotrophic bacteria associated with cold seeps of the Barents Sea. Biogeosciences 9:2947–2960

Haroon MF, Hu S, Shi Y, Imelfort M, Keller J, Hugenholtz P et al (2013) Anaerobic oxidation of methane coupled to nitrate reduction in a novel archaeal lineage. Nature 500:567–570

Harrison BK, Zhang H, Berelson W, Orphan VJ (2009) Variations in archaeal and bacterial diversity associated with the sulfate-methane transition zone in continental margin sediments (Santa Barbara Basin, California). Appl Environ Microbiol 75:1487–1499

Hayashi T, Obata H, Gamo T, Sano Y, Naganuma T (2007) Distribution and phylogenetic characteristics of the genes encoding enzymes relevant to methane oxidation in oxygen minimum zones of the Eastern Pacific Ocean. Res J Environ Sci 1:275–284

Hilário A, Capa M, Dahlgren TG, Halanych KM, Little CTS, Thornhill DJ et al (2011) New perspectives on the ecology and evolution of siboglinid tubeworms. PLoS ONE 6:e16309

Hinrichs K-U, Hayes JM, Sylva SP, Brewer PG, DeLong EF (1999) Methane-consuming archaebacteria in marine sediments. Nature 398:802–805

Holler T, Widdel F, Knittel K, Amann R, Kellermann MY, Hinrichs K-U, Teske A, Boetius A (2011) Thermophilic anaerobic oxidation of methane by marine microbial consortia. ISME J 5:1946–1956

Hu B, Shen L, Lian X, Zhu Q, Liu S, Huang Q et al (2014) Evidence for nitrite-dependent anaerobic methane oxidation as a previously overlooked microbial methane sink in wetlands. Proc Natl Acad Sci U S A 111:4495–4500

Inagaki F, Takai K, Nealson KH, Horikoshi K (2004a) *Sulfurovum lithotrophicum* gen. nov., sp. nov., a novel sulfur-oxidizing chemolithoautotroph within the ε-Proteobacteria isolated from Okinawa Trough hydrothermal sediments. Int J Syst Evol Microbiol 54:1477–1482

Inagaki F, Tsunogai U, Suzuki M, Kosaka A, Machiyama H, Takai K et al (2004b) Characterization of C1-metabolizing prokaryotic communities in methane seep habitats at the Kuroshima Knoll, Southern Ryukyu Arc, by analyzing pmoA, mmoX, mxaF, mcrA, and 16S rRNA genes. Appl Environ Microbiol 70:7445–7455

Jaekel U, Musat N, Adam B, Kuypers M, Grundmann O, Musat F (2013) Anaerobic degradation of propane and butane by sulfate-reducing bacteria enriched from marine hydrocarbon cold seeps. ISME J 7:885–895

Janssen PH, Liesack W, Schink B (2002) *Geovibrio thiophilus* sp. nov., a novel sulfur-reducing bacterium belonging to the phylum Deferribacteres. Int J Syst Evol Microbiol 52:1341–1347

Jones DS, Flood BE, Bailey JV (2015) Metatranscriptomic analysis of diminutive thiomargarita-like bacteria ("Candidatus Thiopilula" spp.) from abyssal cold seeps of the Barbados Accretionary Prism. Appl Environ Microbiol 81:3142–3156

Joye SB, Boetius A, Orcutt BN, Montoya JP, Schulz HN, Erickson MJ, Lugo SK (2004) The anaerobic oxidation of methane and sulfate reduction in sediments from Gulf of Mexico cold seeps. Chem Geol 205:219–238

Kato N, Yurimoto H, Thauer RK (2006) The physiological role of the ribulose monophosphate pathway in bacteria and Archaea. Biosci Biotechnol Biochem 70:10–21

Kirkegaard RH, Dueholm MS, McIlroy SJ, Nierychlo M, Karst SM, Albertsen M, Nielsen PH (2016) Genomic insights into members of the candidate phylum Hyd24-12 common in mesophilic anaerobic digesters. ISME J 10:2352–2364

Kleindienst S, Ramette A, Amann R, Knittel K (2012) Distribution and in situ abundance of sulfate-reducing bacteria in diverse marine hydrocarbon seep sediments. Environ Microbiol 14:2689–2710

Kleindienst S, Herbst F-A, Stagars M, von Netzer F, von Bergen M, Seifert J et al (2014) Diverse sulfate-reducing bacteria of the *Desulfosarcina/Desulfococcus* clade are the key alkane degraders at marine seeps. ISME J 8:2029–2044

Knittel K, Boetius A (2009) Anaerobic oxidation of methane: progress with an unknown process. Annu Rev Microbiol 63:311–334

Knittel K, Boetius A, Lemke A, Eilers H, Lochte K, Pfannkuche O et al (2003) Activity, distribution, and diversity of sulfate reducers and other bacteria in sediments above gas hydrate (Cascadia Margin, Oregon). Geomicrobiol J 20:269–294

Knittel K, Lösekann T, Boetius A, Kort R, Amann R (2005) Diversity and distribution of methanotrophic archaea at cold seeps. Appl Environ Microbiol 71:467–479

Krukenberg V, Harding K, Richter M, Glöckner FO, Gruber-Vodicka HR, Adam B et al (2016) *Candidatus* Desulfofervidus auxilii, a hydrogenotrophic sulfate-reducing bacterium involved in the thermophilic anaerobic oxidation of methane. Environ Microbiol 18:3073–3091

Laso-Pérez R, Wegener G, Knittel K, Widdel F, Harding KJ, Krukenberg V et al (2016) Thermophilic archaea activate butane via alkyl-coenzyme M formation. Nature 539:396–401

Lazar CS, L'Haridon S, Pignet P, Toffin L (2011) Archaeal populations in hypersaline sediments underlying orange microbial mats in the Napoli Mud Volcano. Appl Environ Microbiol 77:3120–3131

Levin LA (2005) Ecology of cold seep sediments: interactions of fauna with flow, chemistry and microbes. In: Gibson RN, Atkinson RJA, Gordon JDM (eds) Oceanography and marine biology: an annual review. Taylor & Francis, Boca Raton, pp 1–46

Li M, Jain S, Baker BJ, Taylor C, Dick GJ (2013) Novel hydrocarbon monooxygenase genes in the metatranscriptome of a natural deep-sea hydrocarbon plume. Environ Microbiol 16:60–71

Lloyd KG, Albert DB, Biddle JF, Chanton JP, Pizarro O, Teske A (2010) Spatial structure and activity of sedimentary microbial communities underlying a *Beggiatoa* spp. mat in a Gulf of Mexico hydrocarbon seep. PLoS One 5:e8738

Lloyd KG, Lapham L, Teske A (2006) An anaerobic methane-oxidizing community of ANME-1b archaea in hypersaline Gulf of Mexico sediments. Appl Environ Microbiol 72:7218–7230

Lösekann T, Knittel K, Nadalig T, Fuchs B, Niemann H, Boetius A, Amann R (2007) Diversity and abundance of aerobic and anaerobic methane oxidizers at the Haakon Mosby Mud Volcano, Barents Sea. Appl Environ Microbiol 73:3348–3362

Marlow JJ, Steele JA, Case DH, Connon SA, Levin LA, Orphan VJ (2014) Microbial abundance and diversity patterns associated with sediments and carbonates from the methane seep environments of Hydrate Ridge, OR. Front Mar Sci 1:44

Martinez-Cruz K, Leewis M-C, Herriott IC, Sepulveda-Jauregui A, Anthony KW, Thalasso F, Leigh MB (2017) Anaerobic oxidation of methane by aerobic methanotrophs in sub-Arctic lake sediments. Sci Total Environ 607–608:23–31

Mastalerz V, de Lange GJ, Dählmann A (2009) Differential aerobic and anaerobic oxidation of hydrocarbon gases discharged at mud volcanoes in the Nile deep-sea fan. Geochim Cosmochim Acta 73:3849–3863

Merkel AY, Huber JA, Chernyh NA, Bonch-Osmolovskaya EA, Lebedinsky AV (2012) Detection of putatively thermophilic anaerobic methanotrophs in diffuse hydrothermal Vent fluids. Appl Environ Microbiol 79:915–923

Meyer S, Wegener G, Lloyd KG, Teske A, Boetius A, Ramette A (2013) Microbial habitat connectivity across spatial scales and hydrothermal temperature gradients at Guaymas Basin. Front Microbiol 4:207

Meyerdierks A, Kube M, Kostadinov I, Teeling H, Glöckner FO, Reinhardt R, Amann R (2010) Metagenome and mRNA expression analyses of anaerobic methanotrophic archaea of the ANME-1 group. Environ Microbiol 12:422–439

Mills HJ, Martinez RJ, Story S, Sobecky PA (2005) Characterization of microbial community structure in Gulf of Mexico gas hydrates: comparative analysis of DNA- and RNA-derived clone libraries. Appl Environ Microbiol 71:3235–3247

Mills HJ, Martinez RJ, Story S, Sobecky PA (2004) Identification of members of the metabolically active microbial populations associated with *Beggiatoa* species mat communities from Gulf of Mexico cold-seep sediments. Appl Environ Microbiol 70:5447–5458

Muyzer G, Stams AJM (2008) The ecology and biotechnology of sulphate-reducing bacteria. Nat Rev Microbiol 6:441–454

Niemann H, Elvert M, Hovland M, Orcutt B, Judd A, Suck I et al (2005) Methane emission and consumption at a North Sea gas seep (Tommeliten area). Biogeosciences 2:335–351

Niemann H, Lösekann T, de Beer D, Elvert M, Nadalig T, Knittel K et al (2006) Novel microbial communities of the Haakon Mosby mud volcano and their role as a methane sink. Nature 443:854–858

Omoregie EO, Mastalerz V, de Lange G, Straub KL, Kappler A, Røy H et al (2008) Biogeochemistry and community composition of iron- and sulfur-precipitating microbial mats at the Chefren Mud Volcano (Nile Deep Sea Fan, Eastern Mediterranean). Appl Environ Microbiol 74:3198–3215

Oni OE, Friedrich MW (2017) Metal oxide reduction linked to anaerobic methane oxidation. Trends Microbiol 25:88–90

Orcutt B, Boetius A, Elvert M, Samarkin V, Joye SB (2005) Molecular biogeochemistry of sulfate reduction, methanogenesis and the anaerobic oxidation of methane at Gulf of Mexico cold seeps. Geochim Cosmochim Acta 69:4267–4281

Orphan VJ, Hinrichs KU, Ussler W, Paull CK, Taylor LT, Sylva SP et al (2001) Comparative analysis of methane-oxidizing archaea and sulfate-reducing bacteria in anoxic marine sediments. Appl Environ Microbiol 67:1922–1934

Orphan VJ, House CH, Hinrichs K-U, McKeegan KD, DeLong EF (2002) Multiple archaeal groups mediate methane oxidation in anoxic cold seep sediments. Proc Natl Acad Sci U S A 99:7663–7668

Oshkin IY, Wegner C-E, Lüke C, Glagolev MV, Filippov IV, Pimenov NV et al (2014) Gammaproteobacterial methanotrophs dominate cold methane seeps in floodplains of West Siberian Rivers. Appl Environ Microbiol 80:5944–5954

Pachiadaki MG, Lykousis V, Stefanou EG, Kormas KA (2010) Prokaryotic community structure and diversity in the sediments of an active submarine mud volcano (Kazan mud volcano, East Mediterranean Sea). FEMS Microbiol Ecol 72:429–444

Paul BG, Ding H, Bagby SC, Kellermann MY, Redmond MC, Andersen GL, Valentine DL (2017) Methane-oxidizing bacteria shunt carbon to microbial mats at a marine hydrocarbon seep. Front Microbiol 8:186

Pernthaler A, Dekas AE, Brown CT, Goffredi SK, Embaye T, Orphan VJ (2008) Diverse syntrophic partnerships from deep-sea methane vents revealed by direct cell capture and metagenomics. Proc Natl Acad Sci U S A 105:7052–7057

Petersen JM, Dubilier N (2009) Methanotrophic symbioses in marine invertebrates. Environ Microbiol Rep 1:319–335

Portillo MC, Leff JW, Lauber CL, Fierer N (2013) Cell size distributions of soil bacterial and archaeal taxa. Appl Environ Microbiol 79:7610–7617

Preisler A, de Beer D, Lichtschlag A, Lavik G, Boetius A, Jørgensen BB (2007) Biological and chemical sulfide oxidation in a *Beggiatoa* inhabited marine sediment. ISME J 1:341–353

Ravenschlag K, Sahm K, Amann R (2001) Quantitative molecular analysis of the microbial community in marine arctic sediments (Svalbard). Appl Environ Microbiol 67:387–395

Redmond MC, Valentine DL, Sessions AL (2010) Identification of novel methane-, ethane-, and propane-oxidizing bacteria at marine hydrocarbon seeps by stable isotope probing. Appl Environ Microbiol 76:6412–6422

Roalkvam I, Jørgensen SL, Chen Y, Stokke R, Dahle H, Hocking WP et al (2011) New insight into stratification of anaerobic methanotrophs in cold seep sediments. FEMS Microbiol Ecol 78:233–243

Roslev P, King GM (1995) Aerobic and anaerobic starvation metabolism in methanotrophic bacteria. Appl Environ Microbiol 61:1563–1570

Roslev P, King GM (1994) Survival and recovery of methanotrophic bacteria starved under oxic and anoxic conditions. Appl Environ Microbiol 60:2602–2608

Rossel PE, Elvert M, Ramette A, Boetius A, Hinrichs K-U (2011) Factors controlling the distribution of anaerobic methanotrophic communities in marine environments: evidence from intact polar membrane lipids. Geochim Cosmochim Acta 75:164–184

Rubin-Blum M, Antony CP, Borowski C, Sayavedra L, Pape T, Sahling H et al (2017) Short-chain alkanes fuel mussel and sponge *Cycloclasticus* symbionts from deep-sea gas and oil seeps. Nat Microbiol 2:17093

Ruff SE, Arnds J, Knittel K, Amann R, Wegener G, Ramette A, Boetius A (2013) Microbial communities of deep-sea methane seeps at Hikurangi Continental Margin (New Zealand). PLoS ONE 8:e72627

Ruff SE, Biddle JF, Teske AP, Knittel K, Boetius A, Ramette A (2015) Global dispersion and local diversification of the methane seep microbiome. Proc Natl Acad Sci U S A 112:4015–4020

Ruff SE, Felden J, Gruber-Vodicka HR, Marcon Y, Knittel K, Ramette A, Boetius A (2019) In situ development of a methanotrophic microbiome in deep-sea sediments. ISME J 13:197–213

Saad S, Bhatnagar S, Tegetmeyer HE, Geelhoed JS, Strous M, Ruff SE (2017) Transient exposure to oxygen or nitrate reveals ecophysiology of fermentative and sulfate-reducing benthic microbial populations. Environ Microbiol 19:4866–4881

Schreiber L, Holler T, Knittel K, Meyerdierks A, Amann R (2010) Identification of the dominant sulfate-reducing bacterial partner of anaerobic methanotrophs of the ANME-2 clade. Environ Microbiol 12:2327–2340

Sender R, Fuchs S, Milo R (2016) Revised estimates for the number of human and bacteria cells in the body. PLoS Biol 14:e1002533

Sivan O, Antler G, Turchyn AV, Marlow JJ, Orphan VJ (2014) Iron oxides stimulate sulfate-driven anaerobic methane oxidation in seeps. Proc Natl Acad Sci U S A 111:E4139–E4147

Sommer S, Linke P, Pfannkuche O, Schleicher T, Schneider von Deimling J, Reitz A et al (2009) Seabed methane emissions and the habitat of frenulate tubeworms on the Captain Arutyunov mud volcano (Gulf of Cadiz). Mar Ecol Prog Ser 382:69–86

Sommer S, Linke P, Pfannkuche O, Niemann H, Treude T (2010) Benthic respiration in a seep habitat dominated by dense beds of ampharetid polychaetes at the Hikurangi Margin (New Zealand). Mar Geol 272:223–232

Stagars MH, Ruff SE, Amann R, Knittel K (2016) High diversity of anaerobic alkane-degrading microbial communities in marine seep sediments based on (1-methylalkyl) succinate synthase genes. Front Microbiol 6:1511

Tavormina PL, Ussler W, Orphan VJ (2008) Planktonic and sediment-associated aerobic methanotrophs in two seep systems along the North American Margin. Appl Environ Microbiol 74:3985–3995

Teske A, Hinrichs K-U, Edgcomb V, de Vera Gomez A, Kysela D, Sylva SP, Sogin ML, Jannasch HW (2002) Microbial diversity of hydrothermal sediments in the Guaymas Basin: evidence for anaerobic methanotrophic communities. Appl Environ Microbiol 68:1994–2007

Thurber AR, Levin LA, Rowden AA, Sommer S, Linke P, Kröger K (2013) Microbes, macrofauna, and methane: a novel seep community fueled by aerobic methanotrophy. Limnol Oceanogr 58:1640–1656

Trembath-Reichert E, Case DH, Orphan VJ (2016) Characterization of microbial associations with methanotrophic archaea and sulfate-reducing bacteria through statistical comparison of nested Magneto-FISH enrichments. PeerJ 4:e1913

Vigneron A, Cruaud P, Pignet P, Caprais J-C, Cambon-Bonavita M-A, Godfroy A, Toffin L (2013) Archaeal and anaerobic methane oxidizer communities in the Sonora Margin cold seeps, Guaymas Basin (Gulf of California). ISME J 7:1595–1608

Vigneron A, Cruaud P, Roussel EG, Pignet P, Caprais J-C, Callac N et al (2014) Phylogenetic and functional diversity of microbial communities associated with subsurface sediments of the Sonora Margin, Guaymas Basin. PLoS One 9:e104427

Vigneron A, Bishop A, Alsop EB, Hull K, Rhodes I, Hendricks R et al (2017) Microbial and isotopic evidence for methane cycling in hydrocarbon-containing groundwater from the Pennsylvania Region. Front Microbiol 8:593

Wankel SD, Adams MM, Johnston DT, Hansel CM, Joye SB, Girguis PR (2012) Anaerobic methane oxidation in metalliferous hydrothermal sediments: influence on carbon flux and decoupling from sulfate reduction. Environ Microbiol 14:2726–2740

Wasmund K, Kurtböke DI, Burns KA, Bourne DG (2009) Microbial diversity in sediments associated with a shallow methane seep in the tropical Timor Sea of Australia reveals a novel aerobic methanotroph diversity. FEMS Microbiol Ecol 68:142–151

Weber HS, Habicht KS, Thamdrup B (2017) Anaerobic methanotrophic archaea of the ANME-2d cluster are active in a low-sulfate, iron-rich freshwater sediment. Front Microbiol 8:619

Wegener G, Krukenberg V, Riedel D, Tegetmeyer HE, Boetius A (2015) Intercellular wiring enables electron transfer between methanotrophic archaea and bacteria. Nature 526:587–590

Wegener G, Shovitri M, Knittel K, Niemann H, Hovland M, Boetius A (2008) Biogeochemical processes and microbial diversity of the Gullfaks and Tommeliten methane seeps (Northern North Sea). Biogeosciences 5:1127–1144

Wegener G, Krukenberg V, Ruff SE, Kellermann MY, Knittel K (2016) Metabolic capabilities of microorganisms involved in and associated with the anaerobic oxidation of methane. Front Microbiol 7:46

Winkel M, Mitzscherling J, Overduin PP, Horn F, Winterfeld M, Rijkers R et al (2018) Anaerobic methanotrophic communities thrive in deep submarine permafrost. Sci Rep 8:1291

Yan T, Ye Q, Zhou J, Zhang CL (2006) Diversity of functional genes for methanotrophs in sediments associated with gas hydrates and hydrocarbon seeps in the Gulf of Mexico. FEMS Microbiol Ecol 57:251–259

Yanagawa K, Sunamura M, Lever MA, Morono Y, Hiruta A, Ishizaki O et al (2011) Niche separation of methanotrophic archaea (ANME-1 and -2) in methane-seep sediments of the Eastern Japan Sea Offshore Joetsu. Geomicrobiol J 28:118–129

Yoshinaga MY, Lazar CS, Elvert M, Lin Y-S, Zhu C, Heuer VB et al (2015) Possible roles of uncultured archaea in carbon cycling in methane-seep sediments. Geochim Cosmochim Acta 164:35–52

Zhang Y, Su X, Chen F, Jiao L, Jiang H, Dong H, Ding G (2012) Abundance and diversity of candidate division JS1- and *Chloroflexi*-related bacteria in cold seep sediments of the northern South China Sea. Front Earth Sci 6:373–382

Zhou Z, Liu Y, Lloyd KG, Pan J, Yang Y, Gu J-D, Li M (2019) Genomic and transcriptomic insights into the ecology and metabolism of benthic archaeal cosmopolitan Thermoprofundales (MBG-D archaea). ISME J 13:885–901

Chapter 2
Anaerobic Hydrocarbon-Degrading Sulfate-Reducing Bacteria at Marine Gas and Oil Seeps

Sara Kleindienst and Katrin Knittel

Abstract Microorganisms are key players in our biosphere because of their ability to degrade various organic compounds including a wide range of hydrocarbons. At hydrocarbon seeps, microorganisms with the ability to utilize diverse hydrocarbons (such as methane, short- and long-chain alkanes, or aromatic hydrocarbons) as carbon and electron source are significantly influencing biogeochemical cycles. Marine hydrocarbon seep sediments are hot spots for microbial activity, particularly for sulfate-reducing bacteria that show elevated respiration rates at these sites. At some seeps, more than 90% of sulfate reduction is potentially coupled to non-methane hydrocarbon oxidation, emphasizing the environmental relevance of these microorganisms and the need to identify key players in situ. Several hydrocarbon-degrading sulfate-reducing bacteria were enriched or isolated from marine sediments, however, in situ active microorganisms were to a large extent represented by uncultivated taxa. Here, we provide an overview of the current understanding of non-methane hydrocarbon-degrading sulfate-reducing bacteria at marine hydrocarbon seeps, including their in situ distribution, abundance, and activity.

Characteristics of Habitats Suitable for Microbial Hydrocarbon Degradation

Prominent habitats of marine hydrocarbon-degrading bacteria and archaea are cold seeps and hot vents (e.g. Boetius and Knittel 2010; Dowell et al. 2016; Widdel et al. 2010). Marine cold seeps are geosystems that are characterized by migration and discharge of hydrocarbons from the subsurface seabed to the hydrosphere. At

S. Kleindienst (✉)
Eberhard Karls University, Tübingen, Germany
e-mail: sara.kleindienst@uni-tuebingen.de

K. Knittel
Max-Planck Institute for Marine Microbiology, Bremen, Germany
e-mail: kknittel@mpi-bremen.de

A. Teske and V. Carvalho (eds.), *Marine Hydrocarbon Seeps*,
Springer Oceanography, https://doi.org/10.1007/978-3-030-34827-4_2

hydrothermal vents of mid ocean ridges, the seafloor consists of basalts and usually lacks a sediment cover. Sedimentary hydrothermal systems such as the Guaymas Basin (Gulf of California) and the Middle Valley (Juan de Fuca Ridge), however, offer suitable habitats within the surface seafloor (Adams et al. 2013; Kallmeyer and Boetius 2004; Laso-Pérez et al. 2016; Teske et al. 2002). Short chain alkanes are formed over relatively short geological time scales via thermogenic processes and often exist at high concentrations. At gas seeps, methane is the dominant hydrocarbon and present in a micromolar to millimolar range besides other short chain alkanes that are typically present as trace gases. Hydrocarbon seeps contain a broad range of alkanes, alkenes and aromatic compounds (Bazylinski et al. 1988; Didyk and Simoneit 1989). Alkenes are absent or only present in traces at marine habitats and are therefore not assumed to be an important energy source for marine microorganisms.

For a long time, it was assumed that the mineralization of hydrocarbons would only be feasible under oxic conditions. This assumption was based on the argument that oxygen is required for the enzymatic activation of hydrocarbons, which are mostly chemically inert (Wilkes and Schwarzbauer 2010). However, since the 1990s, microorganisms have been discovered that are capable of hydrocarbon degradation under anoxic conditions (e.g. Dolfing et al. 1990; Lovley and Lonergan 1990).

In Situ Evidence for Anaerobic Hydrocarbon Degradation by Sulfate-Reducing Bacteria

At hydrocarbon seeps, a large fraction of sulfate reduction is fueled by the anaerobic oxidation of methane (AOM; Reeburgh 2007), but the major part of total sulfate reduction is likely fueled by the oxidation of non-methane hydrocarbons (Bose et al. 2013; Bowles et al. 2011; Kallmeyer and Boetius 2004). At cold seeps in the Gulf of Mexico, in situ biogeochemical indications for microbial non-methane hydrocarbon oxidation came initially from stable carbon isotopic data of gaseous hydrocarbons (Sassen et al. 1999, 2004). The molecular and isotopic compositions of C_2–C_5 hydrocarbons obtained from gas hydrate-bearing sediments below chemosynthetic communities indicated that they were altered by microbial oxidation (Sassen et al. 1999). Furthermore, based on isotopic compositions it was proposed that ethane, isobutane, and isopentane are least affected, while propane, butane, and pentane were most subjected to microbial oxidation (Sassen et al. 2004). In addition, the analysis of carbon isotope compositions of authigenic carbonates demonstrated that methane is a contributor but not the dominant source of metabolic energy at sites of active venting (Formolo et al. 2004). Instead, oxidation of non-methane hydrocarbons was discussed to be the primary source of carbonate alkalinity (Formolo et al. 2004).

At hydrocarbon seeps in the Gulf of Mexico, methane-dependent sulfate reduction rates accounted for less than 10% of total sulfate reduction rates (Joye et al. 2004), indicating that other seep-related compounds such as non-methane hydrocarbons were oxidized by sulfate-reducing bacteria (SRB). Similarly, when rates for AOM and sulfate reduction were measured in hydrothermal vent sediments from Guaymas Basin under different pressure and temperature conditions, AOM contributed to only 1–5% of the sulfate reduction rates (Kallmeyer and Boetius 2004). Furthermore, a global compilation of rate measurements indicated that sulfate reduction rates are enhanced by the presence of aliphatic hydrocarbons or gaseous alkanes (Bowles et al. 2011). Sulfate reduction rates exceeded those of AOM in various seeps, and the estimated average integrated global rate for AOM was determined to be only 5% (Bowles et al. 2011) of previously reported estimates (Hinrichs and Boetius 2002). Another contemporary study estimated rates for anaerobic oxidation of propane in marine environments, and demonstrated the potential importance of this process as a potential sink for propane (Quistad and Valentine 2011). Therefore, SRB may have a greater impact on non-methane hydrocarbon concentrations in the ocean and the atmosphere than previously recognized.

More biogeochemical evidence for anaerobic sulfate-dependent non-methane hydrocarbon degradation came from mud volcanoes in the Gulf of Cadiz (Niemann et al. 2006) as well as from mud volcanoes in the Central Nile deep-sea fan (Mastalerz et al. 2009). The decrease of C_2–C_4 compounds together with the detection of a strong signal of an unresolved complex mixture of hydrocarbons in the sulfate reduction zone indicated that sulfate reduction fueled by higher hydrocarbons could be an important microbial process in mud volcano sediments in the Gulf of Cadiz (Niemann et al. 2006). An unresolved complex mixture typically appears as a hump in the chromatogram of a gas chromatographic analysis, and becomes more defined when oils are biodegraded resulting in a depletion of *n*-alkanes (Niemann et al. 2006). Another study, which investigated molecular and isotopic composition of gaseous hydrocarbons, revealed that sulfate reduction at Isis and Amon mud volcanoes in the Central Nile deep-sea fan was mainly supported by thermogenically-derived methane, propane, and butane (Mastalerz et al. 2009). Ethane and isobutane were also found, but were proposed to be of secondary priority for microorganisms. In addition, distinct overlapping sulfate- and hydrocarbon-rich zones were present, e.g., between 20 and 50 cm below seafloor at the Amon mud volcano (Mastalerz et al. 2009).

The analysis of intact polar membrane lipids, petroleum hydrocarbons, and stable carbon isotopic compositions of hydrocarbon gases from the Chapopote Asphalt Volcano in the southern Gulf of Mexico showed additional evidence for anaerobic hydrocarbon oxidation (see also Chap. 5). Crude oil was found to be biodegraded by lacking, e.g., *n*-alkanes, while diagnostic intact polar membrane lipids indicated that, besides other community members, SRB are present, and may play an important role in petroleum degradation (Schubotz et al. 2011).

Anaerobic Hydrocarbon-Degrading Microorganisms

In the early 1990, the first anaerobic hydrocarbon oxidizers that use toluene with iron or nitrate as electron acceptors have been described (Dolfing et al. 1990; Lovley and Lonergan 1990). Later, also other hydrocarbons were shown to be anaerobically degraded, including short- and long chain alkanes (Aeckersberg et al. 1991; Ehrenreich et al. 2000; Kniemeyer et al. 2007; So and Young 1999; Zengler et al. 1999b), as well as different aromatic compounds (e.g. Fries et al. 1994; Harms et al. 1999a; Meckenstock et al. 1999; Morasch et al. 2004; Rabus et al. 1993; Rabus and Widdel 1995) using nitrate, iron(III), or sulfate as electron acceptors. For an overview see Widdel et al. (2010). Some bacteria also live in syntrophic associations with methanogens (Cheng et al. 2013a, b; Tan et al. 2014; Zengler et al. 1999b) or perform anoxygenic photosynthesis (Zengler et al. 1999a), and some archaea live in syntrophic interactions with SRB (Laso-Pérez et al. 2016; Chen et al. 2019).

Generally, hydrocarbon-oxidizing bacteria belong to the phyla *Actinobacteria, Bacteroidetes, Cyanobacteria, Deinococcus-Thermus, Firmicutes*, and in particular *Proteobacteria*. Within *Archaea, Archaeoglobus fulgidus* (Khelifi et al. 2014), "*Ca.* Syntrophoarchaeum" (Laso-Pérez et al. 2016) and "*Ca.* Argoarchaeum" (Chen et al. 2019) were reported to degrade specific alkanes anaerobically. Based on the detected high diversity of non-canonical methyl-coenzyme M reductases genes (*mcr*) retrieved from metagenomes there are indications for numerous, yet undescribed archaeal branches involved in anaerobic alkane degradation (Dombrowski et al. 2017; Borrel et al. 2019).

Anaerobic Hydrocarbon-Degrading Sulfate-Reducing Bacteria

Hydrocarbon-degrading SRB are highly diverse with respect to substrate usage and their phylogeny (Fig. 1). So far, cultured SRB mainly belong to *Deltaproteobacteria* with the exceptions of *Desulfosporosinus* sp. strain Y5 (Liu et al. 2004) and *Desulfotomaculum* sp. strain OX39 (Morasch et al. 2004) within the *Clostridia*, as well as *Archaeoglobus fulgidus* strain VC-16 (Khelifi et al. 2014). Most deltaproteobacterial isolates were phylogenetically affiliated with the *Desulfosarcina*/*Desulfococcus* clade (Aeckersberg et al. 1991, 1998; Cravo-Laureau et al. 2004; Harms et al. 1999b; Higashioka et al. 2009; Kniemeyer et al. 2007; Meckenstock et al. 2000; So and Young 1999; Watanabe et al. 2017). Additional isolates were affiliated with *Desulfotignum* (Ommedal and Torsvik 2007) and *Desulfatiglans anilini* (Galushko et al. 1999; Harms et al. 1999b; Kniemeyer et al. 2003; Musat et al. 2009; Rabus et al. 1993).

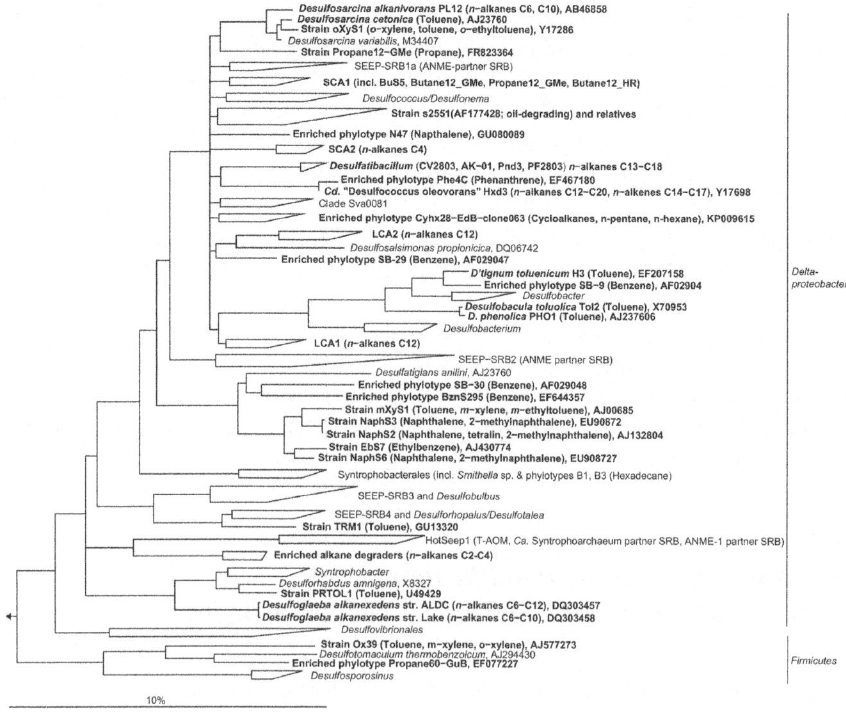

Fig. 1 Phylogenetic tree showing the affiliation of 16S rRNA gene sequences of cultured or enriched sulfate-reducing anaerobic hydrocarbon degraders (printed in boldface) in the context of selected reference sequences of the domain *Bacteria*. Substrate usage of hydrocarbon degraders is given in parentheses. The scale bar represents 10% estimated sequence divergence

So far, isolated SRB were described to degrade either aliphatic hydrocarbons such as C_3–C_{20} *n*-alkanes, C_7–C_{23} *n*-alkenes, or aromatic hydrocarbons such as benzene, toluene, and naphthalene (see Widdel et al. 2010 and references therein). The spectrum of hydrocarbons used by a single microbe, however, is restricted to a narrow range of chain lengths, or even to a single compound. The capability to degrade both, aliphatic and aromatic hydrocarbons, was not yet observed.

So far, only few anaerobic short-chain alkane degraders were enriched and isolated (Jaekel et al. 2013; Kniemeyer et al. 2007). The only cultivated propane and butane degrader is strain BuS5, which was isolated from Guaymas Basin (Kniemeyer et al. 2007). Other enrichments with ethane, propane, and butane, using sediments from hydrothermal systems at the Middle Valley, showed a high 16S rDNA tag sequence abundance of HotSeep1/*Desulfofervidus* affiliated sequences (Adams et al. 2013). *Desulfofervidus auxilii* is the thermophilic sulfate-reducing partner bacterium of "*Ca.* Syntrophoarchaeum", receiving reducing equivalents from the archaeal butane oxidizer (Laso-Pérez et al. 2016).

Until 2016, all known short-chain hydrocarbon degraders were bacteria that couple the complete hydrocarbon oxidization to CO_2 and sulfate reduction in one organism (Adams et al. 2013; Jaekel et al. 2014; Kniemeyer et al. 2003; Musat 2015; Savage et al. 2010). Recently, the thermophilic butane-oxidizing archaea "*Ca.* Syntrophoarchaeum caldarius" and "*Ca.* S. butanivorans" were discovered (Laso-Pérez et al. 2016) that live in syntrophy with their sulfate-reducing partner bacteria of the *Desulfofervidus*/HotSeep1 clade (Krukenberg et al. 2016), forming dense consortia. Yet, it is still unknown which factors select for anaerobic hydrocarbon oxidation by a single bacterial species over consortial, archaeal, and syntrophic hydrocarbon oxidation.

Aromatic hydrocarbon-degrading SRB include strain EbS7, which oxidizes ethylbenzene, isolated from Guaymas Basin (Kniemeyer et al. 2003), and strains NaphS2, NaphS3, and NaphS6, oxidizing e.g. naphthalene and 2-methylnaphthalene, and originating from North Sea and Mediterranean Sea sediments (Galushko et al. 1999; Musat et al. 2009).

Detection and In Situ Quantification of Hydrocarbon-Degrading Sulfate-Reducing Bacteria

Using cultivation-independent approaches, hydrocarbon-oxidizing SRB were investigated across diverse marine hydrocarbon seeps (e.g. Adams et al. 2013; Hassanshahian et al. 2010; Kleindienst et al. 2012, 2014; Orcutt et al. 2010; Quistad and Valentine 2011; Ruff et al. 2016; Savage et al. 2010; Stagars et al. 2017). For instance, the analysis of 16S rRNA gene clone libraries constructed from marine gas and oil seeps in the Gulf of Mexico revealed microbial communities comprising diverse and abundant SRB (Orcutt et al. 2010). In that study, *Deltaproteobacteria* affiliated with members of the uncultivated groups SEEP-SRB3 and SEEP-SRB4, as well as with *Desulfatiglans anilini*, and the *Desulfosarcina*/*Desulfococcus* clade, suggesting that these groups might be involved in or are influenced by the degradation of hydrocarbons or petroleum byproducts (Orcutt et al. 2010). In a parallel study, oily, gas hydrate-rich seep sediments from the Gulf of Mexico (site 161) were analyzed by clone libraries after reverse transcription of rRNA (Kleindienst et al. 2012). Interestingly, the community structure based on RNA showed a different composition than by DNA-based analysis (Fig. 2). RNA-based phylogenetic analyses revealed that *Deltaproteobacteria* clearly dominated all bacteria with about 75% of total sequences (Kleindienst et al. 2012). Of these, the uncultivated group SEEP-SRB2, and the *Desulfosarcina*/*Desulfococcus* clade were found to be relatively high abundant, in contrast to SEEP-SRB3 and SEEP-SRB4 dominating the DNA-based clone libraries. Further in-depth analyses of the *Desulfosarcina*/*Desulfococcus*

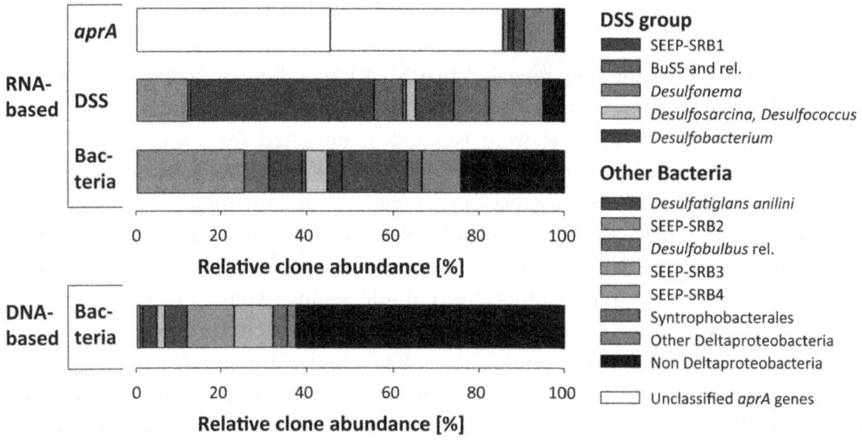

Fig. 2 Diversity of sulfate-reducing microorganisms in Gulf of Mexico sediments (site 161) based on DNA- and RNA-based analyses taken from Kleindienst et al. (2012) and Orcutt et al. (2010). DSS = *Desulfosarcina/Desulfococcus*; *aprA* = adenosine-5′-phosphosulfate reductase subunit A

clade revealed the SEEP-SRB1 group and Bus5-related organisms to be relatively high abundant. RNA-based analysis of a functional gene specific for sulfate-reducing microorganisms, *aprA* (adenosine-5′-phosphosulfate reductase subunit A), indicated two abundant groups that remained unclassified (Fig. 2).

In the same study, the class *Deltaproteobacteria* and the genus *Desulfotomaculum* (*Firmicutes*) were analyzed by CARD-FISH as a proxy for SRB across diverse marine hydrocarbon seep sediments. *Desulfosarcina/Desulfococcus* dominated most of the investigated hydrocarbon seep sediments: up to 53% of total cells (89% of all *Deltaproteobacteria*) at Amon Mud Volcano, up to 36% of total cells (92% of all *Deltaproteobacteria*) at Gulf of Mexico, and even up to 61% of total cells (99% of all *Deltaproteobacteria*) at Hydrate Ridge (Kleindienst et al. 2012). Therefore, the *Desulfosarcina/Desulfococcus* group was assumed to play an important role in hydrocarbon degradation processes. The link between alkane oxidation and specific types of SRB was achieved by stable-isotope probing (SIP) techniques (see also Chap. 10). The key players involved in butane and dodecane degradation in marine gas and hydrocarbon seep sediments were identified as diverse *Desulfosarcina/Desulfococcus* members (Kleindienst et al. 2014), confirming that this dominant group is actually capable of mediating non-methane hydrocarbon oxidation under similar conditions as found in marine seep sediments. These SIP-identified key players were named according to their ability to degrade short-chain alkanes (SCA) or long-chain alkanes (LCA): SCA1, SCA2, LCA1 and LCA2 (Fig. 1).

SCA1

Cultivation-based techniques revealed that SCA1 members are catalyzing propane and butane oxidation. Strain BuS5 and enrichment cultures But12-GMe, Prop12-GMe, and But12-HyR were isolated or enriched from seeps at Guaymas Basin, Gulf of Mexico, and Hydrate Ridge (Jaekel et al. 2013; Kniemeyer et al. 2007; Musat 2015). These propane- and butane-oxidizing isolates and enrichment cultures clustered into a phylogenetically distinct group, named SCA1. The intra-group identity of the 16S rRNA gene sequences was > 94% (Kleindienst 2012), suggesting that SCA1 might represent a single genus. Thus, all SCA1 members may have similar metabolic capabilities with respect to propane and butane degradation. Based on the origin of isolates and enrichment cultures, SCA1 members might be globally distributed at various hydrocarbon seeps including Guaymas Basin, Gulf of Mexico, and Amon Mud Volcano. In addition, the successful enrichment of these bacteria from Hydrate Ridge samples (enrichment culture But12-HyR; Jaekel et al. 2013; Musat 2015) has shown that they also occur in methane-dominated seeps, where they probably use short-chain alkanes that are present in variable, yet often low concentrations (mostly <1%, Milkov et al. 2004). Furthermore, SCA1 sequences were found frequently in 16S rRNA gene clone libraries from seeps in the Gulf of Mexico (Kleindienst et al. 2012; Martinez et al. 2006; Mills et al. 2005), and several Mediterranean mud volcanoes (Pachiadaki et al. 2011). Members of the group SCA1 are therefore suggested as global key players for degradation of short-chain alkanes at marine seeps. After petroleum seepage simulated in a sediment core from the Caspian Sea using an oil-flow-through system, SCA1 cell numbers significantly increased in the top 3 cm (Stagars et al. 2017), supporting the hypothesis about the importance of SCA1 in hydrocarbon degradation.

SCA2

Based on SIP-techniques, dominant butane degraders in Guaymas Basin sediments were affiliated with the SCA2 group (Kleindienst et al. 2014). Group SCA2 contained exclusively the identified butane degraders as well as sequences from butane enrichment cultures that originated from Guaymas Basin sediments, and had a 16S rRNA gene sequence identity of 93% (Kleindienst 2012). Sequences affiliating with SCA2 were reported from other studies and found at diverse habitats including the Gulf of Mexico (Martinez et al. 2006; Mills et al. 2004), several Mediterranean mud volcanoes (Omoregie et al. 2009; Pachiadaki et al. 2010, 2011), Eel River Basin (Orphan et al. 2001; Pernthaler et al. 2008), and non-seep habitats (Acosta-Gonzalez et al. 2013). Nevertheless, SCA2 group members are particularly found at Guaymas Basin, and may therefore be to some extent endemic to certain seeps sites.

LCA1

The LCA1 group comprises dodecane-degrading SRB from the Amon Mud Volcano identified by SIP (Kleindienst et al. 2014), and has an intragroup identity of >94% based on the 16S rRNA gene (Kleindienst 2012). Besides the Amon Mud volcano, LCA1 members were also detected in other habitats, such as coastal sediments (Mediterranean Sea, France), where LCA1 accounted for 35% of deltaproteobacterial sequences (Miralles et al. 2007). Here, in situ experiments indicated that LCA1 members were responsible for hydrocarbon degradation of C_{17}–C_{30} alkanes in sediments amended with oil. Therefore, it can be concluded that members of this group are potential alkane degraders and probably globally distributed, while not only being restricted to seep habitats.

LCA2

LCA2 includes dodecane-degrading organisms from Guaymas Basin as identified by SIP (Kleindienst et al. 2014). Phylogenetic analysis determined a high group similarity of the 16S rRNA genes, with identities >94% (Kleindienst 2012). 16S rRNA gene sequences affiliating to LCA2 were retrieved from seeps at Guaymas Basin (Kleindienst et al. 2014), several submarine volcanoes (Lösekann et al. 2007; Omoregie et al. 2008), Svalbard sediments (Teske et al. 2011), and New Zealand seeps (Ruff et al. 2013). In a sediment-oil-flow-through system set up with a core from the Caspian Sea, LCA2 responded to a simulated petroleum seepage by a 10-fold increase in cell numbers (Stagars et al. 2017). Members of the group LCA2 might all have the same metabolic potential to oxidize long-chain alkanes. However, similar to the group LCA1, it remains to be investigated whether LCA2 organisms are exclusively involved in hydrocarbon degradation, or, if they can also gain their energy from the oxidation of other substrates such as fatty acids.

Desulfatiglans anilini and Relatives

Besides organisms using short-chain and long-chain alkanes as energy sources (e.g. SCA1, SCA2, LCA1, and LCA2), aromatic hydrocarbon degraders such as *Desulfatiglans anilini* and relatives (Fig. 2) are important in marine hydrocarbon seep sediments, which naturally contain high concentrations of aromatic hydrocarbons. Biogeochemical and tracer studies indicated microbial long-chain alkane and aromatic hydrocarbon degradation at seeps from the Gulf of Mexico and Guaymas Basin (Bazylinski et al. 1988; Schubotz et al. 2011). Bacteria capable of degrading aromatic hydrocarbons such as naphthalene, benzene, and toluene were isolated from diverse marine sediments (reviewed by Widdel et al. 2010), including hydrothermal seep sediments in Guaymas Basin (strain EBS7; Kniemeyer et al. 2003). In a global seep study, members of the genus *Desulfatiglans* were particularly abundant in two samples from an oily Gulf of Mexico site (station 161; Kleindienst et al. 2012), while occurring in low numbers or remaining undetected elsewhere. Other surveys have identified 16S rRNA gene sequences of *Desulfatiglans anilini* in various habitats, including seeps on the northern slope of the Gulf of Mexico (Mills et al. 2003; Lloyd et al. 2010), the Tommeliten methane seeps in the North Sea (Wegener et al. 2008), several submarine mud volcanoes (Grünke et al. 2011; Omoregie et al. 2009; Pachiadaki et al. 2010), the methane seeps of the Eel River Basin (Pernthaler et al. 2008), and non-seep habitats such as oil-polluted subtidal sediments after the Prestige oil spill in Spain (Acosta-Gonzalez et al. 2013). *Desulfatiglans anilini* might therefore be important key players involved in the degradation of aromatic hydrocarbons and additional substrates in marine sediments with oil seepage.

Seep-Endemic Sulfate-Reducing Bacteria

Phylogenetic clusters of seep-endemic SRB were identified in various studies by 16S rRNA gene sequence analysis (Fig. 1). These uncultivated groups, named SEEP-SRB (Knittel et al. 2003), were proposed to be involved in hydrocarbon degradation. SEEP-SRB1 was later divided into six subgroups, named SEEP-SRB1a to SEEP-SRB1f (Schreiber et al. 2010). SEEP-SRB1a and SEEP-SRB2 were identified as bacterial partners of ANME-2 archaea at marine seeps, forming microbial consortia involved in the anaerobic oxidation of methane (Kleindienst et al. 2012; Ruff et al. 2016; Wegener et al. 2016; Yanagawa et al. 2013). SEEP-SRB1f is discussed to be involved as partner bacterium of *Ca.* Argoarchaeum in the anaerobic degradation of ethane (Chen et al. 2019). Partial single cell genomes of SEEP-SRB1e and SEEP-SRB1d could not yet confirm their proposed capability to degrade hydrocarbons (Petro et al. 2019). Instead, the authors suggested these groups can use a variety of carbon compounds including complex polysaccharides to maintain a flexible heterotrophic lifestyle. The abundance, distribution, and ecological function of the subgroups SEEP-SRB1b and 1c, as well as of the groups SEEP-SRB3 and SEEP-SRB4 remain still unknown. The main reason for this uncertainty is that the SEEP-SRB groups comprise only uncultured members. SEEP-SRB groups were found to account for up to 2–12% of all free-living cells, i.e. non-aggregated cells, at gas and oil seep sediments as determined by CARD-FISH (Kleindienst et al. 2012), however, their potential involvement in non-methane hydrocarbon oxidation processes remains elusive.

Pathways for Anaerobic Hydrocarbon Degradation

Several pathways for the anaerobic oxidation of hydrocarbons have been described: carboxylation, oxygen-independent hydroxylation, hydration, reversible hydration, and reversal of methanogenesis (Heider and Schühle 2013; Rabus et al. 2016; and references therein). The most intensively studied activation mechanism of non-methane hydrocarbons is the addition to fumarate (Wilkes et al. 2016). This pathway was first described for toluene activation in *Thauera aromatica* strain K172 (Beller and Spormann 1998; Biegert et al. 1996), but was later also found in degradation of alkanes (Kropp et al. 2000; Rabus et al. 2001). The substrates are activated through homolytic C-H bond cleavage at a subterminal carbon atom, and the addition to fumarate, yielding (1-methylalkyl) succinates (Gieg and Suflita 2002; Heider and Schühle 2013; Rabus et al. 2001). Further degradation of the produced alkylsuccinates involves carbon skeleton re-arrangement, de-carboxylation, and β-oxidation. For growth with propane, activation at both, the subterminal or terminal carbon atom, was described (Kniemeyer et al. 2007).

Glycyl radical enzymes are responsible for the fumarate addition. The benzylsuccinate synthase (Bss) is regarded as the prototype enzyme of a large family of fumarate-adding enzymes that are involved in the anaerobic degradation of aromatic and aliphatic hydrocarbons (Heider et al. 2016). While Bss is responsible for toluene degradation (Beller and Spormann 1997; Biegert et al. 1996; Callaghan et al. 2010), (2-naphthylmethyl) succinate synthase (Nms) has been described for naphthalene activation (Annweiler et al. 2000; Musat et al. 2010).

Alkane-activating enzymes were studied in two model organisms: the sulfate-reducing *Desulfatibacillum alkenivorans* AK-01, and the nitrate-reducing *Aromatoleum* HxN1. An alkylsuccinate synthase (Ass) was described for *Desulfatibacillum alkenivorans* AK-01 (Callaghan et al. 2008) and a (1-methyl) alkylsuccinate synthase (Mas) was described for *Aromatoleum* sp. strain HxN1 (Grundmann et al. 2008). Ass and Mas are synonyms and were, thus, used interchangeably. In the thermophilic sulfate reducer *Archaeoglobus fulgidus*, there was evidence for Ass/Mas activity caused by a gene originally annotated as a pyruvate formate lyase (Khelifi et al. 2014). Phylogenetic analysis (Fig. 3a) suggested a bacterial origin for this gene, which was probably acquired from a bacterial donor by lateral gene transfer.

Recent studies by Laso-Pérez et al. (2016) and Chen et al. (2019) suggested a new mechanism for butane and ethane degradation. Archaea of the candidate genera "Syntrophoarchaeum" and "Argoarchaeum" use a modified reverse methanogenesis pathway to degrade these short chain alkanes. They express modified methyl coenzyme M reductases (Mcr) to activate butane and form butyl-CoM (Laso-Pérez et al. 2016) or to activate ethane and form ethyl-CoM (Chen et al. 2019).

Thus, anaerobic archaeal butane activation is possibly analogous to the first step in anaerobic methane oxidation (Krüger et al. 2003; Meyerdierks et al. 2005). In a so far unresolved process butyl-CoM is transformed to butyryl-CoA and subsequently oxidized to acetyl-CoA in the ß-oxidation pathway. Finally, acetyl-CoA is introduced to the reverse methanogenesis pathway and oxidized to CO_2. This process has so far only been shown under thermophilic conditions, and is dependent on their sulfate-reducing partner bacteria of the HotSeep1/*Desulfofervidus* as electron sink (Laso-Pérez et al. 2016). *Ca.* Argoarchaeum does not form aggregates with partner bacteria during ethane oxidation (Chen et al. 2019). These archaea grow together with two types of SEEP-SRB1 but occur mainly as free-living cells.

No nanowires as intercellular connections have been found. As genes known for sulfate reduction have not been detected in these archaea, electrons from ethane degradation have to be transferred to their sulfate-reducing partner bacteria. The physiological transfer mechanism, however, is not yet known.

Genes from butane-oxidizing "*Ca.* Syntrophoarchaeum" encoding modified Mcr are most closely related to those from *Bathyarchaeota* (Evans et al. 2015) and *Ca.* Argoarchaeum, and are clearly different from Mcr retrieved from methanogens and methanotrophs (Fig. 3b). Based on the modified Mcr and other genes found in a metagenomic bin from Guaymas Basin sediments, *Bathyarchaeota* were also suggested to be involved in anaerobic alkane degradation (Dombrowski et al. 2017).

Diversity and Distribution of Fumarate-Adding Marker Genes

Fumarate-adding enzymes contain characteristic conserved motifs including the motif RVXG, which harbors the catalytically essential amino acid glycine (Becker et al. 1999; Grundmann et al. 2008). Due to these conserved motifs, fumarate-adding enzymes are widely used as marker genes for the detection of anaerobic hydrocarbon degraders in the environment (Fig. 3a, for a review see von Netzer et al. 2016). Several primer sets are available, allowing the specific amplification of the catalytic subunit of Bss, Nms, and Ass (e.g. Callaghan et al. 2010; Gittel et al. 2015; von Netzer et al. 2013; Winderl et al. 2007).

Numerous studies were performed studying AssA/MasD in pristine and seepage-impacted sediments from the Mediterranean Sea, North Sea, Caspian Sea, the Pacific and Atlantic Ocean (Acosta-Gonzalez et al. 2013; Callaghan et al. 2010; Gittel et al. 2015; Stagars et al. 2016, 2017; Tan et al. 2015), and in hydrothermal vent systems from Middle Valley (Gittel et al. 2015). BssA/NmsA was mainly studied in contaminated aquifers and soils, as well as in petroleum-contaminated canals and rivers (Callaghan et al. 2010; von Netzer et al. 2013; Winderl et al. 2007; Yagi et al. 2010). All studies revealed a high diversity of fumarate-adding enzymes as shown by the presence of many new phylogenetic clades. Stagars et al. (2016) detected 420 species-level MasD in seven different seep sediments. Due to the limited number of cultivated hydrocarbon degraders and the even more limited number of sequenced fumarate-adding enzymes, there are several clades comprising only sequences from uncultivated organisms, allowing only speculations about their substrate usage and taxonomy.

Seep sites with similar hydrocarbon composition harbored similar alkane-degrading communities suggesting that the MasD community was clearly driven by the hydrocarbon source available (Stagars et al. 2016). Few species-level operational taxonomic units were cosmopolitan and abundant, while ca. 75% were locally restricted (Stagars et al. 2016). Besides few abundant and globally

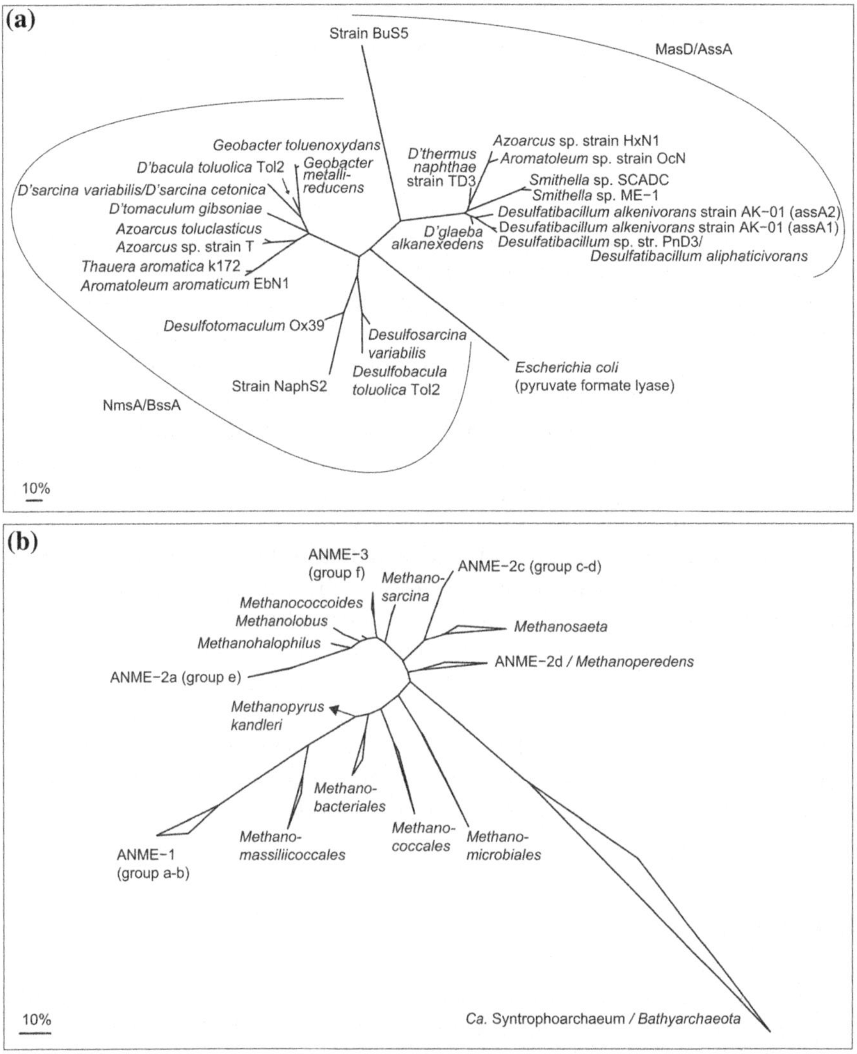

Fig. 3 Phylogenic trees showing **a** the affiliation of hydrocarbon-degrading microorganisms based on the alpha subunit of fumarate-adding enzymes. Bss, benzylsuccinate synthase; Nms, (2-naphthylmethyl) succinate synthase; Ass, alkylsuccinate synthase; Mas, (1-methyl)alkylsuccinate synthase, and **b** the affiliation of the modified methyl coenzyme M reductase subunit alpha (McrA) from butane-degrading "*Ca.* Syntrophoarchaeum" to methanogenic and methanotrophic archaea

distributed alkane degraders there were specialists that have evolved under specific conditions at the diverse seep environments. A rare biosphere of anaerobic alkane degraders was also detected in pristine sediments from the Danish coast (Gittel et al. 2015). These rare bacteria likely increase in abundance in the event of hydrocarbon seepage, or, after anthropogenically caused oil spills, similar to findings for aerobic hydrocarbon degrading bacteria (Kleindienst and Joye 2017).

Outlook

The diversity of SRB capable of hydrocarbon degradation, which are identified as key players at natural hydrocarbon seeps, is steadily increasing. The great diversity of fumarate-adding enzymes retrieved from marine and terrestrial environments, and their phylogenetic position separate from cultivated hydrocarbon degraders, suggests that there are numerous bacterial and/or archaeal lineages of hydrocarbon degraders awaiting their discovery. The recent finding of archaeal short-chain alkane degraders using a modified Mcr for butane and ethane activation that is similar to methane activation by archaeal methanotrophs in AOM allows speculations about possible pathways for other hydrocarbon activation mechanisms, potentially performed by other archaeal lineages.

On one hand, the majority of active sulfate-reducing microorganisms at hydrocarbon seeps are represented by yet uncultivated taxa, and diverse -omic strategies as well as single cell techniques will contribute towards a better understanding of metabolic processes of these organisms. On the other hand, laboratory-based experiments are still crucial to elucidate the physiology of hydrocarbon degraders. Besides metabolic pathways, the putative interaction of hydrocarbon degraders with other microbial community members, niche partitioning, in situ rates of hydrocarbon degradation, and rate-influencing factors are exciting research areas to be addressed in future studies.

References

Acosta-Gonzalez A, Rossello-Mora R, Marques S (2013) Characterization of the anaerobic microbial community in oil-polluted subtidal sediments: aromatic biodegradation potential after the Prestige oil spill. Environ Microbiol 15:77–92

Adams MM, Hoarfrost AL, Bose A, Joye SB, Girguis PR (2013) Anaerobic oxidation of short-chain alkanes in hydrothermal sediments: potential influences on sulfur cycling and microbial diversity. Front Microbiol 4:110

Aeckersberg F, Bak F, Widdel F (1991) Anaerobic oxidation of saturated-hydrocarbons to CO_2 by a new type of sulfate-reducing bacterium. Arch Microbiol 156:5–14

Aeckersberg F, Rainey FA, Widdel F (1998) Growth, natural relationships, cellular fatty acids and metabolic adaptation of sulfate-reducing bacteria that utilize long-chain alkanes under anoxic conditions. Arch Microbiol 170:361–369

Annweiler E, Materna A, Safinowski M, Kappler A, Richnow HH, Michaelis W, Meckenstock RU (2000) Anaerobic degradation of 2-methylnaphthalene by a sulfate-reducing enrichment culture. Appl Environ Microbiol 66:5329–5333

Bazylinski DA, Farrington JW, Jannasch HW (1988) Hydrocarbons in surface sediments from a Guaymas Basin hydrothermal vent site. Org Geochem 12:547–558

Becker A, Fritz-Wolf K, Kabsch W, Knappe J, Schultz S, Volker Wagner AF (1999) Structure and mechanism of the glycyl radical enzyme pyruvate formate-lyase. Nat Struct Biol 6:969–975

Beller HR, Spormann AM (1997) Anaerobic activation of toluene and o-xylene by addition to fumarate in denitrifying strain T. J Bacteriol 179:670–676

Beller HR, Spormann AM (1998) Analysis of the novel benzylsuccinate synthase reaction for anaerobic toluene activation based on structural studies of the product. J Bacteriol 180:5454–5457

Biegert T, Fuchs G, Heider J (1996) Evidence that anaerobic oxidation of toluene in the denitrifying bacterium *Thauera aromatica* is initiated by formation of benzylsuccinate from toluene and fumarate. Eur J Biochem 238:661–668

Boetius A, Knittel K (2010) Habitats of anaerobic methane oxidizers. In: Timmis KN (ed) Handbook of hydrocarbon and lipid microbiology. Springer, Berlin, Heidelberg, pp 2193–2202

Borrel G, Adam PS, McKay LJ, Chen LX, Sierra-García IN, Sieber CMK, Letourneur Q, Ghozlane A, Andersen GL, Li WJ, Hallam SJ, Muyzer G, de Oliveira VM, Inskeep WP, Banfield JF, Gribaldo S (2019) Wide diversity of methane and short-chain alkane metabolisms in uncultured archaea. Nat Microbiol 4:603–613

Bose A, Rogers DR, Adams MM, Joye SB, Girguis PR (2013) Geomicrobiological linkages between short-chain alkane consumption and sulfate reduction rates in seep sediments. Front Microbiol 4:386

Bowles MW, Samarkin VA, Bowles KM, Joye SB (2011) Weak coupling between sulfate reduction and the anaerobic oxidation of methane in methane-rich seafloor sediments during ex situ incubation. Geochim Cosmochim Acta 75:500–519

Callaghan AV, Wawrik B, Ní Chadhain SM, Young LY, Zylstra GJ (2008) Anaerobic alkane-degrading strain AK-01 contains two alkylsuccinate synthase genes. Biochem Biophys Res Commun 366:142–148

Callaghan AV, Davidova IA, Savage-Ashlock K, Parisi VA, Gieg LM, Suflita JM, Kukor JJ, Wawrik B (2010) Diversity of benzyl- and alkylsuccinate synthase genes in hydrocarbon-impacted environments and enrichment cultures. Environ Sci Technol 44:7287–7294

Chen SC, Musat N, Lechtenfeld OJ, Paschke H, Schmidt M, Said N, Popp D, Calabrese F, Stryhanyuk H, Jaekel U, Zhu YG, Joye SB, Richnow HH, Widdel F, Musat F (2019) Anaerobic oxidation of ethane by archaea from a marine hydrocarbon seep. Nature 568:108–111

Cheng L, Ding C, Li Q, He Q, Dai L-r, Zhang H (2013a) DNA-SIP reveals that *Syntrophaceae* play an important role in methanogenic hexadecane degradation. PLoS ONE 8:e66784

Cheng L, Rui J, Li Q, Zhang H, Lu Y (2013b) Enrichment and dynamics of novel syntrophs in a methanogenic hexadecane-degrading culture from a Chinese oilfield. FEMS Microbiol Ecol 83:757–766

Cravo-Laureau C, Matheron R, Joulian C, Cayol JL, Hirschler-Rea A (2004) *Desulfatibacillum alkenivorans* sp. nov., a novel n-alkene-degrading, sulfate-reducing bacterium, and emended description of the genus *Desulfatibacillum*. Int J Syst Evol Microbiol 54:1639–1642

Didyk BM, Simoneit BRT (1989) Hydrothermal oil of Guaymas Basin and implications for petroleum formation mechanisms. Nature 342:65–69

Dolfing J, Zeyer J, Binder-Eicher P, Schwarzenbach RP (1990) Isolation and characterization of a bacterium that mineralizes toluene in the absence of molecular oxygen. Arch Microbiol 154:336–341

Dombrowski N, Seitz KW, Teske AP, Baker BJ (2017) Genomic insights into potential interdependencies in microbial hydrocarbon and nutrient cycling in hydrothermal sediments. Microbiome 5:106

Dowell F, Cardman Z, Dasarathy S, Kellermann MY, Lipp JS, Ruff SE, Biddle JF, McKay LJ, MacGregor BJ, Lloyd KG, Albert DB, Mendlovitz H, Hinrichs KU, Teske A (2016) Microbial communities in methane- and short chain alkane-rich hydrothermal sediments of Guaymas Basin. Front Microbiol 7:17

Ehrenreich P, Behrends A, Harder J, Widdel F (2000) Anaerobic oxidation of alkanes by newly isolated denitrifying bacteria. Arch Microbiol 173:58–64

Evans PN, Parks DH, Chadwick GL, Robbins SJ, Orphan VJ, Golding SD, Tyson GW (2015) Methane metabolism in the archaeal phylum Bathyarchaeota revealed by genome-centric metagenomics. Science 350:434–438

Formolo MJ, Lyons TW, Zhang C, Kelley C, Sassen R, Horita J, Cole DR (2004) Quantifying carbon sources in the formation of authigenic carbonates at gas hydrate sites in the Gulf of Mexico. Chem Geol 205:253–264

Fries MR, Zhou J, Chee-Sanford J, Tiedje JM (1994) Isolation, characterization, and distribution of denitrifying toluene degraders from a variety of habitats. Appl Environ Microbiol 60:2802–2810

Galushko A, Minz D, Schink B, Widdel F (1999) Anaerobic degradation of naphthalene by a pure culture of a novel type of marine sulphate-reducing bacterium. Environ Microbiol 1:415–420

Gieg LM, Suflita JM (2002) Detection of anaerobic metabolites of saturated and aromatic hydrocarbons in petroleum-contaminated aquifers. Environ Sci Technol 36(17):3755–3762

Gittel A, Donhauser J, Røy H, Girguis PR, Jørgensen BB, Kjeldsen KU (2015) Ubiquitous presence and novel diversity of anaerobic alkane degraders in cold marine sediments. Front Microbiol 6:1414

Grundmann O, Behrends A, Rabus R, Amann J, Halder T, Heider J, Widdel F (2008) Genes encoding the candidate enzyme for anaerobic activation of n-alkanes in the denitrifying bacterium, strain HxN1. Environ Microbiol 10:376–385

Grünke S, Felden J, Lichtschlag A, Girnth AC, de Beer D, Wenzhöfer F, Boetius A (2011) Niche differentiation among mat-forming, sulfide-oxidizing bacteria at cold seeps of the Nile Deep Sea Fan (Eastern Mediterranean Sea). Geobiology 9:330–348

Gutierrez T, Kleindienst S (this volume) Uncovering microbial hydrocarbon degradation processes: the promise of stable isotope probing. In: Marine Hydrocarbon Seeps, Springer Oceanography, Teske A, Carvalho V (eds.), Springer, Berlin

Harms G, Rabus R, Widdel F (1999a) Anaerobic oxidation of the aromatic plant hydrocarbon p-cymene by newly isolated denitrifying bacteria. Arch Microbiol 172:303–312

Harms G, Zengler K, Rabus R, Aeckersberg F, Minz D, Rosselló-Móra R, Widdel F (1999b) Anaerobic oxidation of o-xylene, m-xylene, and homologous alkylbenzenes by new types of sulfate-reducing bacteria. Appl Environ Microbiol 65:999–1004

Hassanshahian M, Emtiazi G, Kermanshahi RK, Cappello S (2010) Comparison of oil degrading microbial communities in sediments from the Persian Gulf and Caspian Sea. Soil Sediment Contam: Int J 19:277–291

Heider J, Schühle K (2013) Anaerobic biodegradation of hydrocarbons including methane. In: Rosenberg E, DeLong EF, Lory S, Stackebrandt E, Thompson F (eds) The prokaryotes: prokaryotic physiology and biochemistry. Springer, Berlin, Heidelberg, pp 605–634

Heider J, Szaleniec M, Martins BM, Seyhan D, Buckel W, Golding BT (2016) Structure and function of benzylsuccinate synthase and related fumarate-adding glycyl radical enzymes. J Mol Microbiol Biotechnol 26:29–44

Higashioka Y, Kojima H, Nakagawa T, Sato S, Fukui M (2009) A novel n-alkane-degrading bacterium as a minor member of p-xylene-degrading sulfate-reducing consortium. Biodegradation 20:383–390

Hinrichs K-U, Boetius A (2002) The anaerobic oxidation of methane: new insights in microbial ecology and biogeochemistry. Springer, Berlin

Jaekel U, Musat N, Adam B, Kuypers M, Grundmann O, Musat F (2013) Anaerobic degradation of propane and butane by sulfate-reducing bacteria enriched from marine hydrocarbon cold seeps. ISME J 7:885–895

Jaekel U, Vogt C, Fischer A, Richnow HH, Musat F (2014) Carbon and hydrogen stable isotope fractionation associated with the anaerobic degradation of propane and butane by marine sulfate-reducing bacteria. Environ Microbiol 16:130–140

Joye SB, Boetius A, Orcutt BN, Montoya JP, Schulz HN, Erickson MJ, Lugo SK (2004) The anaerobic oxidation of methane and sulfate reduction in sediments from Gulf of Mexico cold seeps. Chem Geol 205:219–238

Kallmeyer J, Boetius A (2004) Effects of temperature and pressure on sulfate reduction and anaerobic oxidation of methane in hydrothermal sediments of Guaymas Basin. Appl Environ Microbiol 70:1231–1233

Khelifi N, Amin Ali O, Roche P, Grossi V, Brochier-Armanet C, Valette O, Ollivier B, Dolla A, Hirschler-Rea A (2014) Anaerobic oxidation of long-chain n-alkanes by the hyperthermophilic sulfate-reducing archaeon, *Archaeoglobus fulgidus*. ISME J 8:2153–2166

Kleindienst S (2012) Hydrocarbon-degrading sulfate-reducing bacteria in marine hydrocarbon seep sediments. University of Bremen

Kleindienst S, Ramette A, Amann R, Knittel K (2012) Distribution and in situ abundance of sulfate-reducing bacteria in diverse marine hydrocarbon seep sediments. Environ Microbiol 14:2689–2710

Kleindienst S, Herbst F-A, Stagars M, von Netzer F, von Bergen M, Seifert J, Peplies J, Amann R, Musat F, Lueders T, Knittel K (2014) Diverse sulfate-reducing bacteria of the *Desulfosarcina/Desulfococcus* clade are the key alkane degraders at marine seeps. ISME J 8:2029–2044

Kleindienst S, Joye SB (2017) Global aerobic degradation of hydrocarbons in aquatic systems. In: Rojo F (ed) Aerobic utilization of hydrocarbons, oils and lipids. Springer International Publishing, Cham, pp 1–18

Kniemeyer O, Fischer T, Wilkes H, Glöckner FO, Widdel F (2003) Anaerobic degradation of ethylbenzene by a new type of marine sulfate-reducing bacterium. Appl Environ Microbiol 69:760–768

Kniemeyer O, Musat F, Sievert SM, Knittel K, Wilkes H, Blumenberg M, Michaelis W, Classen A, Bolm C, Joye SB, Widdel F (2007) Anaerobic oxidation of short-chain hydrocarbons by marine sulphate-reducing bacteria. Nature 449:898–910

Knittel K, Boetius A, Lemke A, Eilers H, Lochte K, Pfannkuche O, Linke P, Amann R (2003) Activity, distribution, and diversity of sulfate reducers and other bacteria in sediments above gas hydrate (Cascadia margin, Oregon). Geomicrobiol J 20:269–294

Kropp KG, Davidova IA, Suflita JM (2000) Anaerobic oxidation of n-dodecane by an addition reaction in a sulfate-reducing bacterial enrichment culture. Appl Environ Microbiol 66:5393–5398

Krüger M, Meyerdierks A, Glöckner FO, Amann R, Widdel F, Kube M, Reinhardt R, Kahnt J, Bocher R, Thauer RK, Shima S (2003) A conspicuous nickel protein in microbial mats that oxidize methane anaerobically. Nature 426:878–881

Krukenberg V, Harding K, Richter M, Glockner FO, Gruber-Vodicka HR, Adam B, Berg JS, Knittel K, Tegetmeyer HE, Boetius A, Wegener G (2016) *Candidatus* Desulfofervidus auxilii, a hydrogenotrophic sulfate-reducing bacterium involved in the thermophilic anaerobic oxidation of methane. Environ Microbiol 18:3073–3091

Laso-Pérez R, Wegener G, Knittel K, Widdel F, Harding KJ, Krukenberg V, Meier DV, Richter M, Tegetmeyer HE, Riedel D, Richnow HH, Adrian L, Reemtsma T, Lechtenfeld OJ, Musat F (2016) Thermophilic archaea activate butane via alkyl-coenzyme M formation. Nature 539:396–401

Liu A, Garcia-Dominguez E, Rhine ED, Young LY (2004) A novel arsenate respiring isolate that can utilize aromatic substrates. FEMS Microbiol Ecol 48:323–332

Lloyd KG, Albert DB, Biddle JF, Chanton JP, Pizarro O, Teske A (2010) Spatial structure and activity of sedimentary microbial communities underlying a *Beggiatoa* spp. mat in a Gulf of Mexico hydrocarbon seep. PLOS ONE 5:e8738

Lösekann T, Knittel K, Nadalig T, Fuchs B, Niemann H, Boetius A, Amann R (2007) Diversity and abundance of aerobic and anaerobic methane oxidizers at the Haakon Mosby Mud Volcano, Barents Sea. Appl Environ Microbiol 73:3348–3362

Lovley DR, Lonergan DJ (1990) Anaerobic oxidation of toluene, phenol, and p-cresol by the dissimilatory iron-reducing organism, GS-15. Appl Environ Microbiol 56:1858–1864

Martinez RJ, Mills HJ, Story S, Sobecky PA (2006) Prokaryotic diversity and metabolically active microbial populations in sediments from an active mud volcano in the Gulf of Mexico. Environ Microbiol 8:1783–1796

Mastalerz V, de Lange GJ, Dählmann A (2009) Differential aerobic and anaerobic oxidation of hydrocarbon gases discharged at mud volcanoes in the Nile deep-sea fan. Geochim Cosmochim Acta 73:3849–3863

Meckenstock RU, Morasch B, Warthmann R, Schink B, Annweiler E, Michaelis W, Richnow HH (1999) $^{13}C/^{12}C$ isotope fractionation of aromatic hydrocarbons during microbial degradation. Environ Microbiol 1:409–414

Meckenstock RU, Annweiler E, Michaelis W, Richnow HH, Schink B (2000) Anaerobic naphthalene degradation by a sulfate-reducing enrichment culture. Appl Environ Microbiol 66:2743–2747

Meyerdierks A, Kube M, Lombardot T, Knittel K, Bauer M, Glöckner FO, Reinhardt R, Amann R (2005) Insights into the genomes of archaea mediating the anaerobic oxidation of methane. Environ Microbiol 7(12):1937–1951

Milkov AV, Claypool GE, Lee Y-J, Torres ME, Borowski WS, Tomaru H, Sassen R, Long PE (2004) Ethane enrichment and propane depletion in subsurface gases indicate gas hydrate occurrence in marine sediments at southern Hydrate Ridge offshore Oregon. Org Geochem 35:1067–1080

Mills HJ, Hodges C, Wilson K, MacDonald IR, Sobecky PA (2003) Microbial diversity in sediments associated with surface-breaching gas hydrate mounds in the Gulf of Mexico. FEMS Microbiol Ecol 46:39–52

Mills HJ, Martinez RJ, Story S, Sobecky PA (2004) Identification of members of the metabolically active microbial populations associated with *Beggiatoa* species mat communities from Gulf of Mexico cold seep sediments. Appl Environ Microbiol 70(9):5447–5458

Mills HJ, Martinez RJ, Story S, Sobecky PA (2005) Characterization of microbial community structure in Gulf of Mexico gas hydrates: comparative analysis of DNA- and RNA-derived clone libraries. Appl Environ Microbiol 71:3235–3247

Miralles G, Grossi V, Acquaviva M, Duran R, Bertrand JC, Cuny P (2007) Alkane biodegradation and dynamics of phylogenetic subgroups of sulfate-reducing bacteria in an anoxic coastal marine sediment artificially contaminated with oil. Chemosphere 68:1327–1334

Morasch B, Schink B, Tebbe CC, Meckenstock RU (2004) Degradation of o-xylene and m-xylene by a novel sulfate-reducer belonging to the genus *Desulfotomaculum*. Arch Microbiol 181:407–417

Musat F, Galushko A, Jacob J, Widdel F, Kube M, Reinhardt R, Wilkes H, Schink B, Rabus R (2009) Anaerobic degradation of naphthalene and 2-methylnaphthalene by strains of marine sulfate-reducing bacteria. Environ Microbiol 11:209–219

Musat F, Wilkes H, Behrends A, Woebken D, Widdel F (2010) Microbial nitrate-dependent cyclohexane degradation coupled with anaerobic ammonium oxidation. ISME J 4:1290–1301

Musat F (2015) The anaerobic degradation of gaseous, nonmethane alkanes—from in situ processes to microorganisms. Comput Struct Biotechnol J 13:222–228

Niemann H, Duarte J, Hensen C, Omoregie E, Magalhães VH, Elvert M, Pinheiro LM, Kopf A, Boetius A (2006) Microbial methane turnover at mud volcanoes of the Gulf of Cadiz. Geochim Cosmochim Acta 70:5336–5355

Ommedal H, Torsvik T (2007) *Desulfotignum toluenicum* sp. nov., a novel toluene-degrading, sulphate-reducing bacterium isolated from an oil-reservoir model column. Int J Syst Evol Microbiol 57:2865–2869

Omoregie EO, Mastalerz V, de Lange G, Straub KL, Kappler A, Røy H, Stadnitskaia A, Foucher JP, Boetius A (2008) Biogeochemistry and community composition of iron- and sulfur-precipitating microbial mats at the Chefren mud volcano (Nile Deep Sea fan, Eastern Mediterranean). Appl Environ Microbiol 74:3198–3215

Omoregie EO, Niemann H, Mastalerz V, de Lange GJ, Stadnitskaia A, Mascle J, Foucher JP, Boetius A (2009) Microbial methane oxidation and sulfate reduction at cold seeps of the deep Eastern Mediterranean Sea. Mar Geology 261:114–127

Orcutt BN, Joye SB, Kleindienst S, Knittel K, Ramette A, Reitz A, Samarkin V, Treude T, Boetius A (2010) Impact of natural oil and higher hydrocarbons on microbial diversity, distribution, and activity in Gulf of Mexico cold-seep sediments. Deep-Sea Res Part II 57:2008–2021

Orphan VJ, Hinrichs K-U, Ussler W, Paull CK, Taylor LT, Sylva SP, Hayes JM, DeLong EF (2001) Comparative analysis of methane-oxidizing archaea and sulfate-reducing bacteria in anoxic marine sediments. Appl Environ Microbiol 67:1922–1934

Pachiadaki MG, Lykousis V, Stefanou EG, Kormas KA (2010) Prokaryotic community structure and diversity in the sediments of an active submarine mud volcano (Kazan mud volcano, East Mediterranean Sea). FEMS Microbiol Ecol 72:429–444

Pachiadaki M, Kallionaki A, Dählmann A, De Lange G, Kormas K (2011) Diversity and spatial distribution of prokaryotic communities along a sediment vertical profile of a deep-sea mud volcano. Microb Ecol 62:655–668

Pernthaler A, Dekas AE, Brown CT, Goffredi SK, Embaye T, Orphan VJ (2008) Diverse syntrophic partnerships from deep-sea methane vents revealed by direct cell capture and metagenomics. Proc Natl Acad Sci U S A 105:7052–7057

Petro C, Jochum LM, Schreiber L, Marshall IPG, Schramm A, Kjeldsen KU (2019) Single-cell amplified genomes of two uncultivated members of the deltaproteobacterial SEEP-SRB1 clade, isolated from marine sediment. Mar Genomics 46:66–69

Quistad SD, Valentine DL (2011) Anaerobic propane oxidation in marine hydrocarbon seep sediments. Geochim Cosmochim Acta 75:2159–2169

Rabus R, Nordhaus R, Ludwig W, Widdel F (1993) Complete oxidation of toluene under strictly anoxic conditions by a new sulfate-reducing bacterium. Appl Environ Microbiol 59:1444–1451

Rabus R, Widdel F (1995) Anaerobic degradation of ethylbenzene and other aromatic-hydrocarbons by new denitrifying bacteria. Arch Microbiol 163:96–103

Rabus R, Wilkes H, Behrends A, Armstroff A, Fischer T, Pierik AJ, Widdel F (2001) Anaerobic initial reaction of n-alkanes in a denitrifying bacterium: evidence for (1-methylpentyl)succinate as initial product and for involvement of an organic radical in n-hexane metabolism. J Bacteriol 183:1707–1715

Rabus R, Boll M, Heider J, Meckenstock RU, Buckel W, Einsle O, Ermler U, Golding B, Gunsalus R, Kroneck P, Krüger M, Lueders T, Martins B, Musat F, Richnow H, Schink B, Seifert J, Szaleniec M, Treude T, Ullmann G, Vogt C, von Bergen M, Wilkes H (2016) Anaerobic microbial degradation of hydrocarbons: from enzymatic reactions to the environment. J Mol Microbiol Biotechnol 26:5–28

Reeburgh WS (2007) Oceanic methane biogeochemistry. Chem Rev 107:486–513

Ruff SE, Arnds J, Knittel K, Amann R, Wegener G, Ramette A, Boetius A (2013) Microbial communities of deep-sea methane seeps at Hikurangi continental margin (New Zealand). PLoS ONE 8:e72627

Ruff SE, Kuhfuss H, Wegener G, Lott C, Ramette A, Wiedling J, Knittel K, Weber M (2016) Methane seep in shallow-water permeable sediment harbors high diversity of anaerobic methanotrophic communities, Elba, Italy. Front Microbiol 7:374

Sassen R, Joye SB, Sweet ST, DeFreitas DA, Milkov AV, MacDonald IR (1999) Thermogenic gas hydrates and hydrocarbon gases in complex chemosynthetic communities, Gulf of Mexico continental slope. Org Geochem 30:485–497

Sassen R, Roberts HH, Carney R, Milkov AV, DeFreitas DA, Lanoil B, Zhang C (2004) Free hydrocarbon gas, gas hydrate, and authigenic minerals in chemosynthetic communities of the northern Gulf of Mexico continental slope: relation to microbial processes. Chem Geol 205:195–217

Savage KN, Krumholz LR, Gieg LM, Parisi VA, Suflita JM, Allen J, Philp RP, Elshahed MS (2010) Biodegradation of low-molecular-weight alkanes under mesophilic, sulfate-reducing conditions: metabolic intermediates and community patterns. FEMS Microbiol Ecol 72:485–495

Schreiber L, Holler T, Knittel K, Meyerdierks A, Amann R (2010) Identification of the dominant sulfate-reducing bacterial partner of anaerobic methanotrophs of the ANME-2 clade. Environ Microbiol 12:2327–2340

Schubotz F, Lipp JS, Elvert M, Kasten S, Mollar XP, Zabel M, Bohrmann G, Hinrichs K-U (2011) Petroleum degradation and associated microbial signatures at the Chapopote asphalt volcano, Southern Gulf of Mexico. Geochim Cosmochim Acta 75:4377–4398

So CM, Young LY (1999) Isolation and characterization of a sulfate-reducing bacterium that anaerobically degrades alkanes. Appl Environ Microbiol 65:2969–2976

Stagars MH, Ruff SE, Amann R, Knittel K (2016) High diversity of anaerobic alkane-degrading microbial communities in marine seep sediments based on (1-methylalkyl)succinate synthase genes. Front Microbiol 6:1511

Stagars MH, Mishra S, Treude T, Amann R, Knittel K (2017) Microbial community response to simulated petroleum seepage in Caspian Sea sediments. Front Microbiol 8:764

Tan B, Nesbo C, Foght J (2014) Re-analysis of omics data indicates *Smithella* may degrade alkanes by addition to fumarate under methanogenic conditions. ISME J 8:2353–2356

Tan B, Fowler SJ, Abu Laban N, Dong X, Sensen CW, Foght J, Gieg LM (2015) Comparative analysis of metagenomes from three methanogenic hydrocarbon-degrading enrichment cultures with 41 environmental samples. ISME J 9:2028–2045

Teske A, Hinrichs K-U, Edgcomb V, de Vera Gomez A, Kysela D, Sylva SP, Sogin ML, Jannasch HW (2002) Microbial diversity of hydrothermal sediments in the Guaymas Basin: evidence for anaerobic methanotrophic communities. Appl Environ Microbiol 68:1994–2007

Teske A, Durbin A, Ziervogel K, Cox C, Arnosti C (2011) Microbial community composition and function in permanently cold seawater and sediments from an arctic fjord of Svalbard. Appl Environ Microbiol 77:2008–2018

von Netzer F, Pilloni G, Kleindienst S, Kruger M, Knittel K, Grundger F, Lueders T (2013) Enhanced gene detection assays for fumarate-adding enzymes allow uncovering of anaerobic hydrocarbon degraders in terrestrial and marine systems. Appl Environ Microbiol 79:543–552

von Netzer F, Kuntze K, Vogt C, Richnow HH, Boll M, Lueders T (2016) Functional gene markers for fumarate-adding and dearomatizing key enzymes in anaerobic aromatic hydrocarbon degradation in terrestrial environments. J Mol Microbiol Biotechnol 26:180–194

Watanabe M, Higashioka Y, Kojima H, Fukui M (2017) *Desulfosarcina widdelii* sp. nov. and *Desulfosarcina alkanivorans* sp. nov., hydrocarbon-degrading sulfate-reducing bacteria isolated from marine sediment and emended description of the genus *Desulfosarcina*. Int J Syst Evol Microbiol 67:2994–2997

Wegener G, Shovitri M, Knittel K, Niemann H, Hovland M, Boetius A (2008) Biogeochemical processes and microbial diversity of the Gullfaks and Tommeliten methane seeps (Northern North Sea). Biogeosciences 5:1127–1144

Wegener G, Krukenberg V, Ruff SE, Kellermann MY, Knittel K (2016) Metabolic capabilities of microorganisms involved in and associated with the anaerobic oxidation of methane. Front Microbiol 7:46

Widdel F, Knittel K, Galushko A (2010) Anaerobic hydrocarbon-degrading microorganisms: an overview. In: Timmis KN, McGenity T, van der Meer JR, de Lorenzo V (eds) Handbook of hydrocarbon and lipid microbiology. Springer, Berlin, Heidelberg, pp 1997–2021

Wilkes H, Schwarzbauer J (2010) Hydrocarbons: an introduction to structure, physico-chemical properties and natural occurrence. In: Timmis KN, McGenity T, van der Meer JR, de Lorenzo V (eds) Handbook of hydrocarbon and lipid microbiology. Springer, Berlin, Heidelberg, pp 5–48

Wilkes H, Buckel W, Golding BT, Rabus R (2016) Metabolism of hydrocarbons in n-alkane-utilizing anaerobic bacteria. J Mol Microbiol Biotechnol 26:138–151

Winderl C, Schaefer S, Lueders T (2007) Detection of anaerobic toluene and hydrocarbon degraders in contaminated aquifers using benzylsuccinate synthase (bssA) genes as a functional marker. Environ Microbiol 9:1035–1046

Yagi JM, Suflita JM, Gieg LM, DeRito CM, Jeon C-O, Madsen EL (2010) Subsurface cycling of nitrogen and anaerobic aromatic hydrocarbon biodegradation revealed by nucleic acid and metabolic biomarkers. Appl Environ Microbiol 76:3124–3134

Yanagawa K, Morono Y, de Beer D, Haeckel M, Sunamura M, Futagami T, Hoshino T, Terada T, Nakamura K-i, Urabe T, Rehder G, Boetius A, Inagaki F (2013) Metabolically active microbial communities in marine sediment under high-CO_2 and low-pH extremes. ISME J 7:555–567

Zengler K, Heider J, Rosselló-Móra R, Widdel F (1999a) Phototrophic utilization of toluene under anoxic conditions by a new strain of *Blastochloris sulfoviridis*. Arch Microbiol 172:204–212

Zengler K, Richnow HH, Rosselló-Móra R, Michaelis W, Widdel F (1999b) Methane formation from long-chain alkanes by anaerobic microorganisms. Nature 401:266–269

Chapter 3
Guaymas Basin, a Hydrothermal Hydrocarbon Seep Ecosystem

Andreas Teske

Abstract The hydrothermal sediments of Guaymas Basin in the Gulf of California combine several microbial ecosystems. Since Guaymas Basin is an active hydrothermal spreading center, it sustains chemosynthetic microbial communities with inorganic electron donors such as sulfide, hydrogen, ammonia, and reduced metals within steep thermal gradients of hydrothermal sediments and chimneys. At the same time, Guaymas Basin is an organic-rich continental margin site, where high sedimentation rates resulting from high phytoplankton productivity and terrestrial runoff produce organic-rich sediments that support abundant heterotrophic microbial populations. Most interesting in the context of hydrocarbon seepage, hydrothermal heating of organic-rich sediments creates hot hydrocarbon seeps in Guaymas Basin. Aliphatic and aromatic hydrocarbons are generated under high temperature and pressure in the subsurface, and migrate to the sediment surface where they are assimilated and oxidized by hydrocarbon-oxidizing bacteria and archaea. This complex habitat mosaic results in unusually diverse microbial communities with new phylogenetic lineages and surprising physiological capabilities. This chapter provides some background and highlights recent microbial discoveries in Guaymas Basin.

General Characteristics of Guaymas Basin

The Guaymas Basin in the Gulf of California (Fig. 1) is a young marginal rift basin characterized by active seafloor spreading and rapid deposition of organic-rich sediments from highly productive overlying waters, supplemented locally by terrigenous sedimentation from the Sonora Margin (Calvert 1966). It provides the classic example for young spreading centers that are often thickly sedimented due to their proximity to terrigenous sediment sources and the influence of coastal

A. Teske (✉)
University of North Carolina at Chapel Hill, Chapel Hill, NC, USA
e-mail: teske@email.unc.edu

A. Teske and V. Carvalho (eds.), *Marine Hydrocarbon Seeps*,
Springer Oceanography, https://doi.org/10.1007/978-3-030-34827-4_3

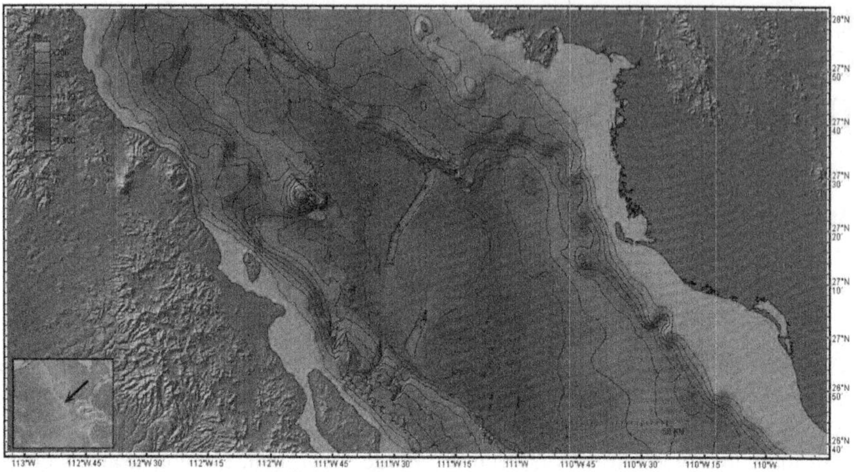

Fig. 1 Guaymas Basin bathymetric map. The northern and southern Guaymas troughs are visible in the center of the map. They are running orthogonal to the northwest/southeast extent of the Gulf of California, and reach a maximum depth just below 2000 m; they are surrounded by extensive gently sloping flanking regions of 1600–1800 m depth, and by the steep slopes of Baja California and the Sonora Margin. The most frequently sampled and well-studied hydrothermal area is located in the southern Guaymas trough. The depth contours correspond to 200 m intervals (Courtesy of Carlos Mortera, UNAM, Institute of Geophysics)

upwelling. The northern and southern axial troughs of Guaymas Basin are bounded by extensive axial-parallel fault lines on both sides (Lonsdale and Becker 1985; Fisher and Becker 1991).

Organic-rich sediments of several hundred meters thickness overlie the spreading centers of Guaymas Basin, and alternate with shallow intrusions of doleritic sills into the unconsolidated sediments (Einsele et al. 1980). While unsedimented mid-ocean ridges show focused magma emplacement directly at the spreading center, the sediments in Guaymas Basin act as a thermal blanket that allows broader off-axis magma emplacement into the sedimented flanking regions (Lizarralde et al. 2011; Berndt et al. 2016). Emplacement of hot magmatic sills indurates and hydrothermally alters their surrounding sediment matrix (Fig. 2); while sills are gradually cooling, they continue to shape hydrothermal circulation patterns and reaction pathways (Saunders et al. 1982; Gieskes et al. 1982; Kastner 1982).

The juxtaposition of active seafloor spreading, sill emplacement, and thick sediment layers has created a dynamic environment where physical, chemical, and biological processes regulate the cycling of sedimentary carbon. Most significantly, buried organic matter in the Guaymas sediments is heated at 200–300 °C under high pressure and transformed quickly into hydrocarbons. Guaymas Basin hydrocarbons are young enough to be ^{14}C-dated and have an average radiocarbon age of approximately 5000 years (Peter et al. 1991). Hydrothermal pyrolysis transforms and mobilizes a major proportion of subsurface carbon sources: the organic carbon

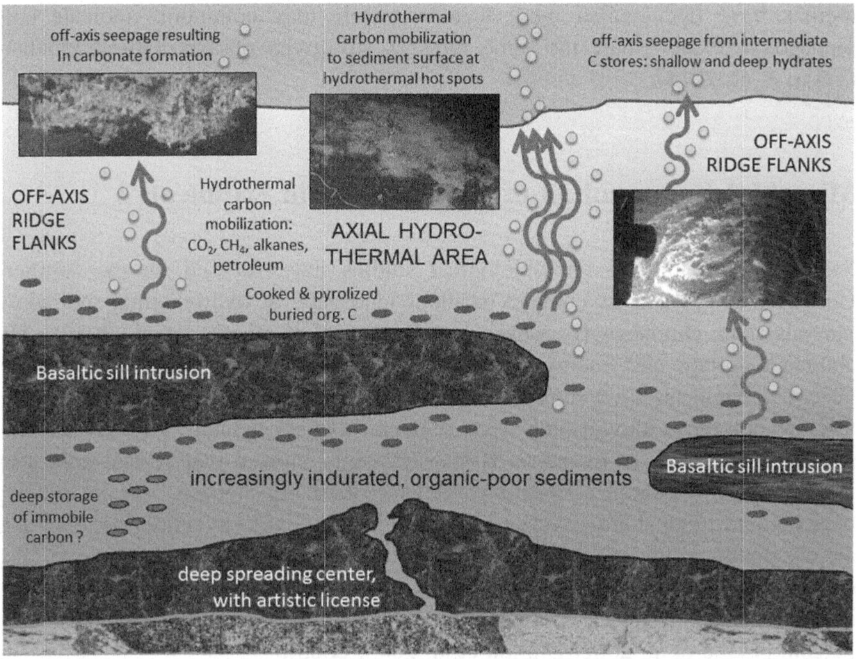

Fig. 2 Concept sketch of hydrothermal flow and transport at Guaymas Basin, on-axis and off-axis. Hydrothermal structures, high-temperature sediments, and microbial mats are found in the axial center that is characterized by vigorous hydrothermal circulation and shallow, hot, recently emplaced sills (Teske et al. 2016). In contrast, seafloor carbonates and methane hydrates occur in cool, off-axis locations, and are sustained by residual seepage from cooling sills in flanking region sediments (Lizarralde et al. 2011)

content of approx. 3–4 wt% in surficial Guaymas Basin sediments (De la Lanza-Espino and Soto 1999) is reduced to 1–2% in subsurface sediments below sills (Rullkötter et al. 1982; Simoneit and Bode 1982). Hydrothermal alteration of sedimentary organic matter generates petroleum compounds including complex mixtures of linear, branched, and cycloalkanes, hopanes, steranes, diasteranes, olefins, and polynuclear aromatic hydrocarbons (Simoneit and Lonsdale 1982; Kawka and Simoneit 1987; Didyk and Simoneit 1989), low-molecular weight alkanes (Bazylinski et al. 1988), organic acids (Martens 1990), and ammonia (Von Damm et al. 1985). These substrates can be used by microbial communities in hydrothermal seafloor sediments and deep subsurface sediments (Teske et al. 2014). Methane seepage also extends towards off-axis locations, where buried, gradually cooling sills extend sufficiently close to the sediment surface to drive gas flow and residual hydrothermal circulation (Lizarralde et al. 2011). During R/V *Atlantis* cruise AT37-06 in December 2016, submersible observations of abundant seafloor mineral mounds and shallow gas hydrates on the northwestern Guaymas flanking regions have documented these features: while they commonly indicate cold seepage, here they share the same roots as sill-driven on-axis hydrothermalism (Teske et al. 2018).

Microbial Communities of Guaymas Basin Sediments

Seafloor sampling and microbiological studies have focused on the southern Guaymas trough and its complex seafloor landscape of hydrothermal sediments, mounds, and chimneys (Lonsdale and Becker 1985; Teske et al. 2016). The abundant carbon and energy sources in hydrothermal fluids and sediments of Guaymas Basin sustain complex and diverse communities of *Bacteria* and *Archaea* that are otherwise known only from separate, distinct habitats. Already the first sequencing surveys of Guaymas Basin sediments showed that typical cold seep inhabitants, for example, anaerobic methane-oxidizing archaea, sulfate-reducing bacteria, and large sulfide-oxidizing bacteria thrive next to hyperthermophiles that are characteristic for mid-ocean ridge hydrothermal vents. These communities also coexist with benthic bacteria and archaea that are common in organic-rich seafloor sediments and subsurface sediments without hydrocarbon seepage or hydrothermalism (Teske et al. 2002; Dhillon et al. 2003, 2005).

Sulfur-Oxidizing Mat-Forming Bacteria in Guaymas Basin

Conspicuous, white, yellow and orange mats of sulfur-oxidizing, nitrate-reducing filamentous *Beggiatoaceae* (Jannasch et al. 1989; Nelson et al. 1989) occur on the surface of hydrothermally active, methane- and sulfide-rich sediments in Guaymas Basin. The *Beggiatoaceae* mats of Guaymas Basin are providing a well-studied model system to investigate the ecophysiology of these organisms. A complementary overview of *Beggiatoaceae* from Gulf of Mexico hydrocarbon seeps is given in Chap. 8, and previously unpublished results on Guaymas *Beggiatoaceae* are presented in Chap. 9. The Guaymas mats can grow to a thickness of several centimeters, and may resemble fluffy pillows that often surround the tubes and gills of the tube worm *Riftia pachyptila* (Fig. 3a, b), which is a keystone species of marine invertebrates commonly encountered at eastern Pacific hydrothermal vents. *Riftia* tube worms require convective mixing of sulfidic hydrothermal fluids and oxygenated seawater in order to provide both compounds simultaneously to their chemosynthetic sulfur-oxidizing endosymbionts (Luther et al. 2001); these conditions most likely favor massive *Beggiatoaceae* accumulations as well.

Since many large sulfur-oxidizing bacteria are sensitive to fully oxidized conditions, the reduced oxygen content of the bottom water in Guaymas Basin

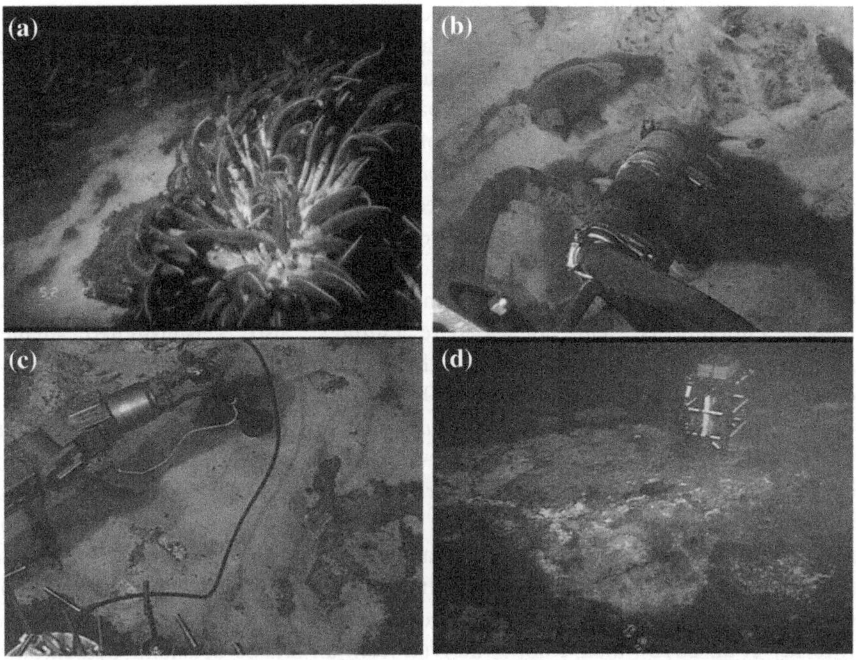

Fig. 3 Microbial mats in Guaymas Basin. **a** Thick *Beggiatoaceae* mats surrounding a cluster of the chemosynthetic tube worm *Riftia pachyptila*, immediately conspicuous by its red gills that take up oxygen, sulfide, and DIC to sustain its chemosynthetic endobionts. **b** Suction sampling of pillow-like *Beggiatoaceae* mats growing between *Riftia* on the surface of a hydrothermal mound (Mat Mound; Teske et al. 2016, and Dowell et al. 2016). The tubeworms are almost completely embedded in fur-like biomass of filamentous sulfur oxidizers. **c** Temperature profiling in an orange and white *Beggiatoaceae* mat, using Alvin's heatflow probe. **d** The "Megamat" area, a large and extremely hot hydrothermal hot spot in Guaymas Basin, covered predominantly with highly textured sulfur precipitates, but mostly lacking the smooth, pillow-like mats of *Beggiatoaceae*. A distinct section of Megamat with *Beggiatoaceae* mats extends to the left of the picture frame (McKay et al. 2012). Image (**a**): collection of Holger W. Jannasch, WHOI. Images (**b–d**): Alvin frame grabber images obtained during dives 4484, 4568, and 4486. Courtesy of the Woods Hole Oceanographic Institution

(ca. 10–20% of seawater saturation; Calvert 1964) favors the development of these abundant *Beggiatoaceae* mats. As already noted at the discovery of the Guaymas *Beggiatoaceae* mats (Jannasch et al. 1989), and substantiated with in situ micro-electrode measurements of their highly dynamic sulfide and oxygen gradients (Gundersen et al. 1992; Teske et al. 2016), pillow-like mats in Guaymas Basin contrast with the classical morphology of sulfur-oxidizing bacterial mats, where the filaments cluster tightly at the diffusion-controlled interface where steep opposing gradients of electron donor (sulfide) and electron acceptor (oxygen or nitrate) overlap on a scale of micrometers (Møller et al. 1985; Nelson et al. 1986).

While current studies have not covered the full diversity of Guaymas *Beggiatoaceae* mats, they indicate that the architecture of the *Beggiatoaceae* mats and their distinct populations are intricately linked to the habitat preferences of these large sulfur-oxidizing bacteria, based on hydrothermal activity and flux (McKay et al. 2012). The central area of hydrothermal hot spots is preferentially colonized by orange-pigmented *Beggiatoaceae*, with a filament diameter of ca. 35–40 μm, whereas the periphery is populated by large, non-pigmented *Beggiatoaceae* with filament diameters around and above 100 μm (Fig. 4a–e). The resulting "fried egg" pattern of orange and white mats on hydrothermal hot spots is common at Guaymas Basin (Fig. 3c). These patterns should be understood as the result of natural enrichments, and not as pure cultures; also, yet-unstudied smaller *Beggiatoaceae* filaments occur among these predominant populations in various degrees of admixture (Nelson et al. 1989).

Although both dominant *Beggiatoaceae* types thrive in cool temperatures of ca. 10 °C at the sediment-seawater interface (McKay et al. 2012), their habitat preferences are connected to differences in underlying hydrothermal flux of electron donors, for example sulfide, and carbon substrates such as methane-derived, hydrothermal, or marine DIC, or low-molecular weight organic acids. The orange-pigmented filaments preferentially colonize high-flux sediments, where high concentrations of hydrothermal electron and carbon sources appear closest to the sediment surface (McKay et al. 2012); their color is derived from an orange octaheme cytochrome, quantitatively the dominant mat protein, with multiple in vitro functionalities including nitrite reduction (MacGregor et al. 2013a). The near-complete genome of the orange filaments indicates versatile oxidative and reductive S cycle pathways, and the ability to assimilate and metabolize organic and inorganic C sources (MacGregor et al. 2013b, c). Attempts to identify the dominant carbon source by stable carbon isotopic analysis of filament biomass have produced interesting preliminary results; the analysis of individual 120 μm-wide white filaments via secondary ion mass spectrometry finds $\delta^{13}C$ values averaging at −42‰, with a standard deviation of ±23‰. Despite this high deviation from the mean, a comparison with other isotopic studies of Guaymas Basin microbial biomass and carbon sources (Pearson et al. 2005) suggests that these large white filaments assimilate $\delta^{13}C$-depleted substrates that are ultimately derived from, or impacted by, methane carbon (McKay et al. 2012).

Microbial Nitrogen Cycling in Surficial Sediments and Mats of Guaymas Basin

In addition to nitrogen-fixing ANME archaea (studied at cold seeps by Pernthaler et al. 2008; Dekas et al. 2009; Miyazaki et al. 2009), the extensive Guaymas *Beggiatoaceae* mats most likely constitute a crucial interface where microbial carbon utilization intersects with N and S cycles. Collaborative studies of in situ chemical and temperature profiles in orange *Beggiatoaceae* mats using microprofiler landers (Dirk de Beer, Max Planck Institute for Marine Microbiology, Bremen,

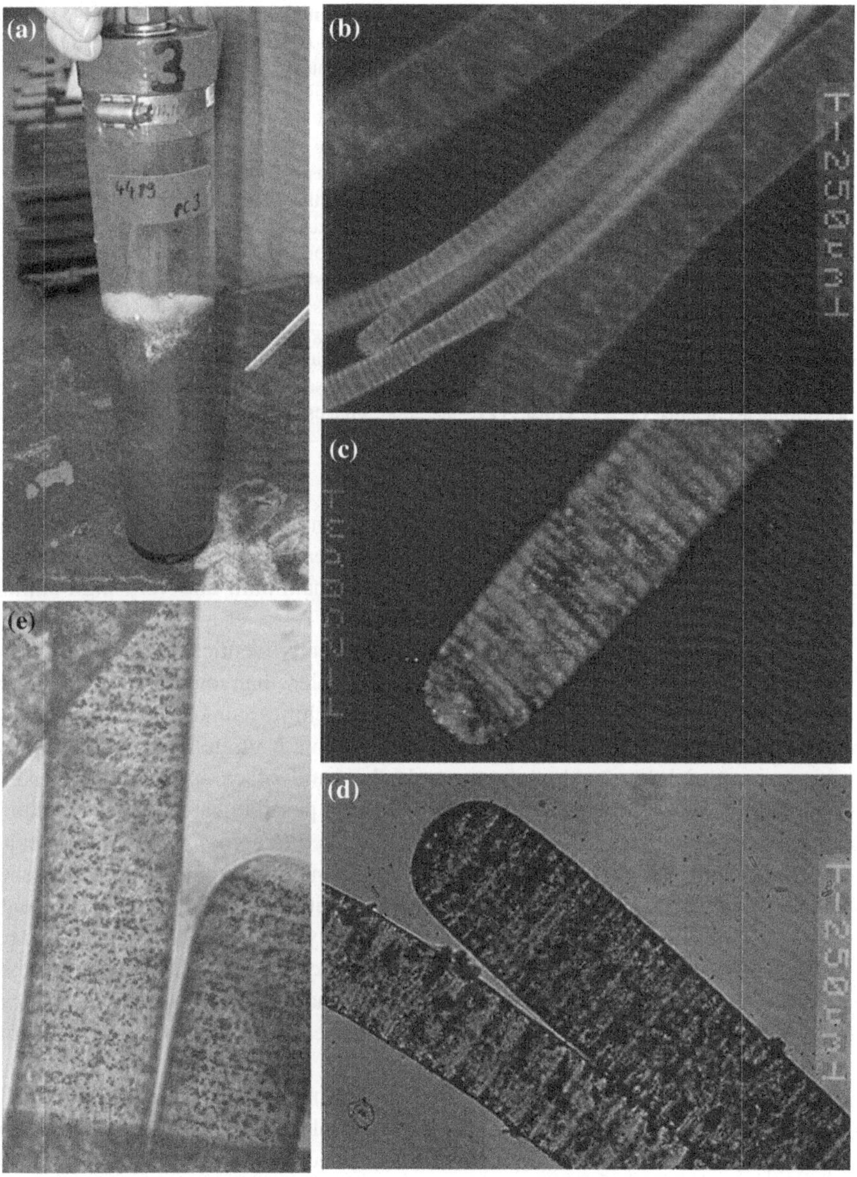
(a)
(b)
(c)
(d)
(e)

◂**Fig. 4 Large sulfur-oxidizing filaments from hydrothermal hot spots in Guaymas Basin.** **a** Freshly collected sediment core from Alvin dive 4489 (December 13, 2008) with a thick *Beggiatoaceae* mat on top. Visually, the mat is dominated by large, white *Beggiatoaceae* filaments. **b** UV Epifluorescence image of two different types of Guaymas Basin *Beggiatoaceae*; the smaller filaments (ca. 35–40 μm in diameter) with yellow fluorescence are the dominant types observed in orange-pigmented *Beggiatoaceae* mats at the center of hydrothermal hot spots (McKay et al. 2012). The larger filaments with blue epifluorescence represent the white *Beggiatoaceae* filaments (ca. 100–120 μm in diameter) that dominate on the perimeter of hydrothermal hot spots, surrounding the orange mat in the center (McKay et al. 2012); the white types can form extensive mats by themselves, for example on the slopes of Mat Mound (Dowell et al. 2016). **c** Darkfield microphotograph of large white *Beggiatoaceae* filaments; the discontinuous distribution of bright sulfur globules indicates individual cells in the multicellular filament. **d** Transmitted light microphotograph of large *Beggiatoaceae*, surrounded by swarming regular-sized bacteria that are visible as small dots. **e** Image of large, white Guaymas Basin *Beggiatoaceae* taken through a dissecting microscope (Cruise AT15-40, December 2008). Image (**a**): Shipboard Science Crew of cruise AT15-40, 2008; Images (**b–d**): A. Teske, UNC Chapel Hill; Image (**e**): J.F. Biddle, University of Delaware

Germany) revealed steep concentration gradients indicative of dynamic geochemical and microbial interactions: a steep oxygen decrease and a nitrate/nitrite increase at the mat interface are followed by a local pH maximum and very high sulfide concentrations immediately beneath the interface (Teske et al. 2016). The in situ temperature at the interface, ca. 3–4 °C, was practically identical to the temperature of Guaymas Basin bottom water, somewhat cooler than measured with Alvin temperature probes, but, reinforcing the psychrophilic habitat preference of the *Beggiatoaceae* mats (McKay et al. 2012). Combined nitrate and nitrite concentrations increased from a bottom water background of ca. 20 μM (most likely dominated by nitrate) to ca. 75 μM within the mat, possibly caused by intracellular nitrate accumulation and leakage by large, vacuolated *Beggiatoaceae* (McKay et al. 2012), as well as nitrifying activity by ammonia-oxidizing, nitrite-producing archaea (*Thaumarchaeota*) that grow associated with the *Beggiatoaceae* filaments (Winkel et al. 2014). Oxygen was generally quickly consumed at the mat surface, while narrow local oxygen peaks below the sediment surface may indicate advective transport, for example, by hydrothermal pumping that re-introduces pockets of oxygenated seawater into shallow sediments near hydrothermal hot spots (Gundersen et al. 1992; Teske et al. 2016).

Nitrate, the principal electron acceptor of the mat-forming *Beggiatoaceae*, is used differently by different types. The near-complete genome of the orange filament type lacked conventional denitrification pathways beyond the reduction of nitrate to nitrite, and alternative pathways were suggested based on the observation that the dominant orange cytochrome of this organism shows nitrite reductase activity (MacGregor et al. 2013b, c). Ex situ incubation experiments with ^{15}N-labelled nitrate showed that the orange *Beggiatoaceae* reduce nitrate to ammonia but do not perform denitrification (Schutte et al. 2018). Prompted by uncertainty about the physiological nitrite reduction pathway in the orange *Beggiatoaceae*, gene expression experiments with two additional candidate nitrite reductases from this organism (in addition to the orange cytochrome) have identified a multifunctional enzyme with nitrite reductase and tetrathionate reductase activity, apparently capable of using partially reduced sulfur species as electron acceptor (Buckley et al. 2019). This capability is reminiscent of elemental sulfur reduction observed in freshwater hot spring *Beggiatoa* sp. (Nelson and Castenholz 1981) and may constitute an adaptation to pulses of reducing conditions in hydrothermal hot spots. In contrast, the white *Beggiatoaceae* at the mat periphery encode the complete pathways for denitrification and for reduction of nitrate to ammonia in their genome; microelectrode profiling on reconstituted mat samples in the lab showed that these filaments toggle between nitrate reduction to dinitrogen in the surficial, nitrate-replete mat layers, and ammonia production in deeper mat layers, which is favored under nitrate-limiting and sulfide-replete conditions (Schutte et al. 2018). Thus, the organism is capable of adapting its nitrate reduction pathways to shifting environmental conditions in the highly dynamic Guaymas Basin hydrothermal sediments. The genomic and physiological differences of the two *Beggiatoaceae* types are reflected by 16S rRNA phylogeny (McKay et al. 2012); although both types form sister lineages within the radiation of the gammaproteobacterial *Beggiatoaceae*, they are sufficiently distant from each other to require future classification as separate genera (Teske and Salman 2014).

Benthic nitrate consumption and dinitrogen production by denitrification are not limited to *Beggiatoaceae* mats. High denitrification rates were measured ex situ in hydrothermally active sediments populated with *Beggiatoaceae* mats, and also in adjacent oil-stained bare sediments (Bowles et al. 2012); these rates decreased significantly in the presence of increased sulfide concentrations, as observed generally in estuarine and marine sediments. Based on 16S rRNA gene surveys, potentially denitrifying candidate organisms included members of the genus *Sulfurimonas* (*Epsilonproteobacteria*), moderately thermophilic denitrifying bacteria that are capable of autotrophic growth with sulfide- and hydrogen as electron donors (Campbell et al. 2006). Functional gene surveys using the nitrous oxide reductase gene (*nosZ*) yielded uncultured phylotypes affiliated with members of the *Alphaproteobacteria* (Bowles et al. 2012). The sediments of Guaymas Basin are rich in ammonia, detected in concentrations of ca. 10–15 mM in hydrothermal fluids (Von Damm et al. 1985), and ca. 2 mM in sediment porewater (Russ et al. 2013). Ammonia coexisting with nitrite, produced by nitrifying archaea or generated as the first intermediate of nitrate reduction, fuels anaerobic ammonia oxidation (anammox), the alternate pathway of nitrogen loss under strictly anaerobic conditions (Mulder et al. 1995). Sequence signatures of marine anammox bacteria (genus *Scalindua*) were indeed found in Guaymas Basin sediments (Russ et al. 2013). To summarize, these studies have identified several microbial key players of the nitrogen cycle in Guaymas Basin: thaumarchaeotal ammonia oxidizers, *Beggiatoaceae, Sulfurimonas*, and uncultured *Alphaproteobacteria* as denitrifyers, *Beggiatoaceae* as ammonia-producing nitrate reducers, and anammox bacteria of the genus *Scalindua*. Together, these microbial populations indicate coexisting oxidative and reductive nitrogen transformations in hydrothermal sediments of Guaymas Basin.

Microbial Cycling of Methane and Other Hydrocarbons in Guaymas Basin

Microbial Communities of Guaymas Basin Hydrothermal Mounds, and Chimneys

The microbial communities of hydrothermal mounds and chimneys in Guaymas Basin differ from those in sediments (Fig. 5). When freshly formed hydrothermal mineral deposits (dominated by calcite, barite, anhydrite, and metal sulfides) coupled with in situ temperature monitoring using thermocouple arrays (Fig. 5a) were analyzed for microbial colonization by DNA extraction and 16S rRNA gene sequencing, mostly hyperthermophilic, anaerobic archaea were found, such as *Methanocaldococcus* spp., other *Methanococcaceae, Korarchaeota, Methanosarcinales*, several uncultured deep-sea hydrothermal vent euryarchaeotal groups, and the *Desulfurococcaceae*. Members of the thaumarchaeotal 'Marine Group I' were the only archaea of obvious seawater origin in these samples (Pagé et al. 2008). These archaeal communities shifted from autotrophic, CO_2/H_2-dependent hyperthermophilic methanogens (predominantly *Methanocaldococcus* sp.) that were detected on freshly deposited, 4-day old chimney materials, towards methylotrophic and/or acetoclastic methanogens (*Methanosarcinales*) and fermentative heterotrophic thermophiles (*Korarchaeota, Aciduliprofundales*) in older chimney material collected after 72 days (Pagé et al. 2008). Most likely, this community shift responds to the accumulation and increasing availability of biomass for heterotrophic metabolism.

Interestingly, when a hydrocarbon-soaked hydrothermal chimney was examined by metagenomic sequencing, thermophilic archaeal families (*Archaeoglobaceae* and *Thermococcaceae*) were among the most frequently detected, followed by sulfate-reducing *Deltaproteobacteria, Thermotogaceae*, and diverse sulfur-oxidizing or sulfate-reducing bacterial families (He et al. 2013). Functional key genes related to those of aromatics- and alkane-degrading sulfate-reducing bacteria, such as *Desulfoglaeba, Desulfotomaculum, Desulfatibacillum*, and *Desulfococcus*, could be detected as well (He et al. 2013). Thus, some bacterial community signatures that are usually known from oily sediments can be recovered from hydrothermal chimneys as well, as long as suitable hydrocarbon substrates are available.

Methanogens and Methane Oxidizers in Guaymas Basin

When hydrocarbon-rich fluids migrate to the upper sediment column they provide fossil carbon substrates to highly active, benthic microbial communities that oxidize and assimilate them (Pearson et al. 2005). Thus, hydrocarbon-rich Guaymas Basin has particular promise for in-depth investigations of hydrocarbon transformations, and the diversity and evolution of hydrocarbon-degrading microorganisms and

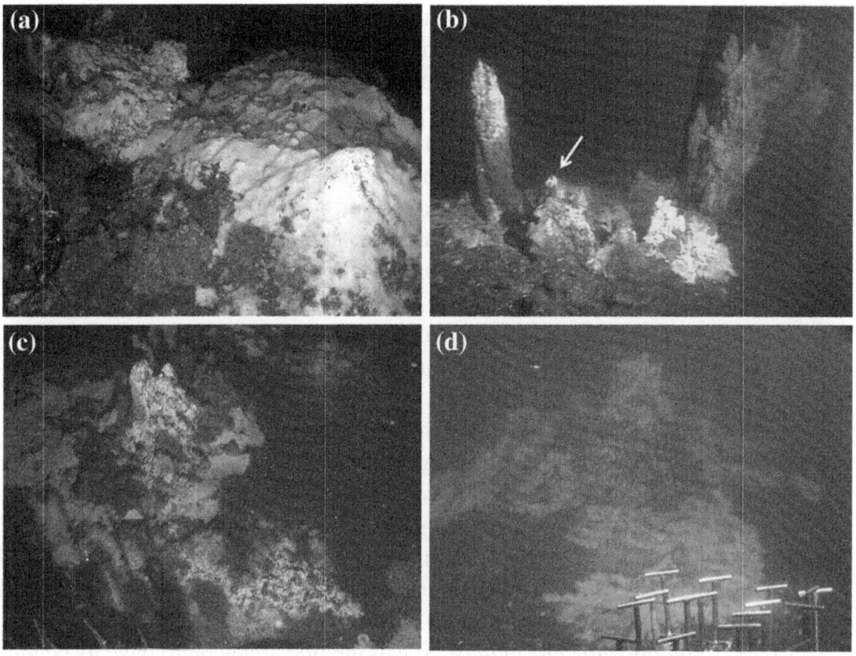

Fig. 5 Hydrothermal mounds and chimneys on Guaymas Basin sediments. **a** Mineral deposits with active hydrothermal venting on the left, visible by shimmering water; the mineral precipitates to the right are overgrown by a *Beggiatoaceae* mat that highlights their rough surface texture. At this location, the formation of new mineral precipitates and its microbial colonization was studied in detail, and a new colonization device is visible in the lower left corner (Pagé et al. 2008; for further context, see Fig. 12B in Teske et al. 2016). **b** Photomosaic of hydrothermal chimneys with white sulfur precipitates, visited during Alvin dive 4492. The arrow points at a small, actively venting chimney that had its top broken off by Alvin; the temperature of the mostly clear, shimmering vent fluid emerging from this orifice was measured as 206 °C (A. Ramette, Report of Alvin Dive 4492). Nearby larger chimneys are partially covered by white sulfur precipitates and by microbial mats. **c** Alvin external still photograph of a complex landscape of microbial mats and sulfur crusts centered on a small hydrothermal mound, ca. 0.5 m high; the mound is covered with bright sulfur precipitates. Immediately to the left of the mound is the instrumented "Marker 14" mat, a thermally, microbiologically and biogeochemically well-characterized microbial mat site showing hydrothermally controlled microbial zonation (McKay et al. 2016). **d** Mat Mound, a large hydrothermal mound, ca. 2.5 m high, is covered with *Riftia* and sulfur-oxidizing mats, and is surrounded by slopes of hydrothermal sediment. The site has been extensively characterized by thermal profiles, geochemical, and microbiological analyses (Dowell et al. 2016). Images (**a**, **c**, **d**) are external still photographs of HOV Alvin; Image (**b**) is a composite of frame grabber images from Dive 4492. Courtesy of the Woods Hole Oceanographic Institution

pathways. Methane is one of the dominant carbon compounds in the hydrothermal sediments of Guaymas Basin (Welhan 1988), and the microbiological exploration of Guaymas Basin has emphasized methanogenesis and methane oxidation at high temperatures. Guaymas Basin provides a model system where the microbial oxidation or assimilation of methane could attenuate the hydrothermally catalyzed mobilization and loss of buried carbon from hydrothermal sediments, a process with strong climate history relevance (Lizarralde et al. 2011). Interestingly, methanogenesis was among the first microbial processes documented in the deep sedimentary subsurface of Guaymas Basin. Leg 64 of the Deep-Sea Drilling Program (DSDP) examined the extent of subsurface life in the massive sediments of Guaymas Basin (Curray et al. 1979; Curray and Moore 1982) and demonstrated microbial methanogenesis in the deep sediment column (Galimov and Simoneit 1982; Oremland et al. 1982). Molecular detection of methanogens and ANME archaea in subsurface sediments of Guaymas Basin should be the next step. Recently, 16S rRNA sequences of ANME-1 archaea have been found in off-axis hydrothermal locations where subsurface methane is seeping towards the sediment surface (Teske et al. 2019).

Methanogens in Guaymas Basin sediments include numerous hyperthermophilic lineages. The most thermophilic methanogen that is currently known, *Methanopyrus kandleri*, was isolated from Guaymas Basin sediments. This hyperthermophilic, H_2/CO_2-dependent species and genus has a growth limit at 110 °C (Kurr et al. 1991), and can survive temperatures of 122 °C under deep-sea in situ pressure (Takai et al. 2008). The Guaymas Basin isolate *Methanocaldococcus jannaschii* and related strains are growing at temperatures of 80–90 °C (Jones et al. 1983, 1989; Jeanthon et al. 1999). *Methanocaldoccus jannaschii* was the first archaeon to have its genome sequenced, demonstrating that the archaea were profoundly distinct from the bacteria not just by 16S rRNA gene sequencing, but on the level of the entire genome (Bult et al. 1996). As shown by a cloning and sequencing survey based on the key gene of methanogenesis (methyl coenzymeA reductase alpha subunit, or *mcrA*), the diversity of methanogens extends beyond these cultured representatives: at least three additional, currently uncultured family-level lineages have been detected in hydrothermal sediments of Guaymas Basin (Lever and Teske 2015). Phylogenetically divergent types of *mcrA* genes that are not covered by conventional PCR primers extend the diversity of methanogens (not only in Guaymas Basin, compare McKay et al. 2017), and are implicated in the activation, assimilation and oxidation of other short-chain alkanes beyond methane (Dombrowski et al. 2017, 2018).

An unusual candidate methanogen from Guaymas Basin was proposed based on metagenomic reconstruction of its genome. The potential methanogen group "Methanofastidiosa" appears to specialize in methylthiol usage as the sole methanogenic substrate and entirely lacks the reductive branch of the Wood-Ljungdahl pathway for the reduction of CO_2 to the cofactor-bound methyl group (Nobu et al. 2016). The "Methanofastidiosa" are phylogenetically congruent with the "Euryarchaeota Guaymas Group", recovered repeatedly in PCR-based 16S rRNA gene sequencing surveys from Guaymas Basin sediments (Teske et al. 2002; Dhillon et al. 2005). Members of the Methanofastidiosa are widely distributed in anaerobic sediments and bioreactors, and do not require particular hydrothermal conditions; instead, the availability of hydrogen and methylthiols may be the key factor for the environmental occurrence and distribution of this metabolically restricted group of yet-uncultured methanogenic candidates (Nobu et al. 2016).

The hydrothermal sediments of Guaymas Basin support sulfate-dependent anaerobic oxidation of methane (AOM) (Fig. 6). This process is usually catalyzed by multiple types of anaerobic methanotrophic archaea (ANME-1, ANME- 2, and ANME-3), acting in syntrophic partnership with specific sulfate-reducing bacteria (*Deltaproteobacteria*), resulting in morphologically complex cell consortia (Knittel and Boetius 2009; AOM occurrence also outside of Guaymas Basin is elaborated further in Chap. 1). Within consortia of ANME-2 archaea and sulfate-reducing bacteria, electrons are transferred from the methane-oxidizing archaea to their sulfate-reducing partners not by diffusion of a reduced substrate, but directly via multi-haem cytochromes produced by ANMEs, which are embedded in the cellular membranes of both partners and also occur in the interstitial space between the cells (McGlynn et al. 2015).

While ANME-2 archaea and their consortia are widespread at cold seeps, sequencing surveys have recovered mainly ANME-1 archaea, including new lineages within this group, in hydrothermal sediments of Guaymas Basin (Dowell et al. 2016). The ANME-2 archaea exist in Guaymas Basin as well, but are mostly limited to surficial sediments with moderate temperatures (Fig. 6). Initial evidence for anaerobic methane oxidation in Guaymas Basin sediments came from 16S rRNA gene sequences of ANME archaea and compound-specific δ^{13}C-signatures of archaeal lipids (Teske et al. 2002), and from activity measurements in ex situ laboratory incubations at high temperatures (Kallmeyer and Boetius 2004). Ex situ laboratory measurements of anaerobic, sulfate-dependent methane oxidation by ANME-1 archaea showed consistent and robust activity at warm temperatures of 37 °C (Kellermann et al. 2012), and in the thermophilic range at 55–60 °C (Holler et al. 2011). Also, high-temperature tolerance up to ca. 75 °C is possible (Holler et al. 2011; Wegener et al. 2015). This wide range of thermal tolerance in ANME-1 archaea is most likely selected for as a consequence of strong thermal fluctuations, due to hydrothermal circulation and seawater entrainment in hot spot sediments (McKay et al. 2016). Congruent with this working hypothesis, phylotypes of recently identified Guaymas-specific phylogenetic clusters within ANME-1 (termed ANME-1a Guaymas I and II; Holler et al. 2011) and of the ANME-1 sister lineage ANME-1Guaymas (Biddle et al. 2012) were recovered in both hot and cold sediments of Guaymas Basin (Fig. 6). The specific ex situ thermal limits of these different ANME-1 subgroups remain to be determined by group-specific enrichment and cultivation.

By the same token, the capability of ANME archaea for nitrogen assimilation requires further investigation. Studies at cold seeps have shown that ANME-1 archaea are capable of nitrogen fixation (Pernthaler et al. 2008; Dekas et al. 2009; Miyazaki et al. 2009). ANME-1 archaea in Guaymas Basin would not require this capability since ammonia occurs in millimolar porewater concentrations in Guaymas Basin hydrothermal sediments (Schutte et al. 2018), but the extent to which Guaymas ANME archaea have adapted to these special conditions is at present unknown.

FISH hybridization of ANME-1 consortia from Guaymas Basin revealed associations with unusual bacterial syntrophs—the HotSeep-1 group—in high-temperature

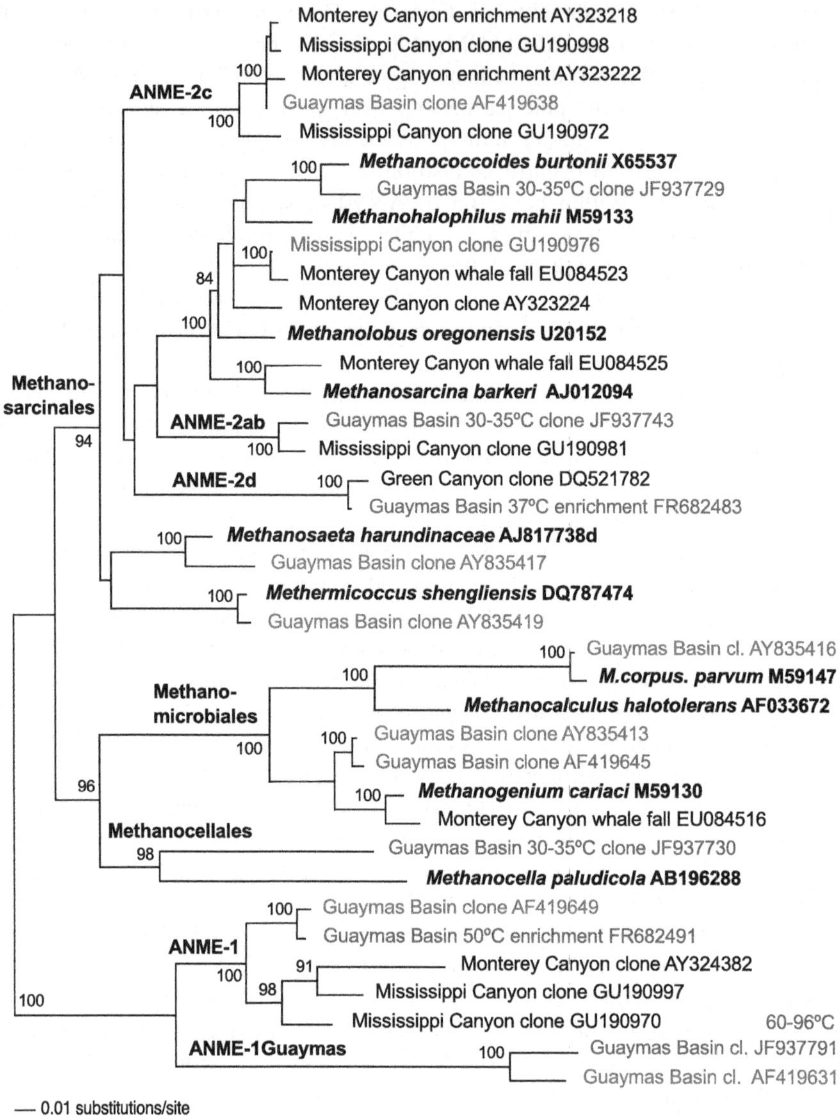

Fig. 6 Methanogens and ANME Archaea in Guaymas Basin. Phylogenetic tree of methanogens and methane-oxidizing archaea from sediment samples with different thermal regimes in Guaymas Basin, compared to cold seep sediments in Monterey Canyon (CA), cold methane hydrate site Mississippi Canyon 118 in the northern Gulf of Mexico, and pure cultures of methanogens. This small selection from currently available 16S rRNA gene phylotypes illustrates overlapping phylotypes among the ANME-1 and ANME-2 archaea from hot (red) and cold (blue) sediments, contrasting with distinct, Guaymas-specific phylotypes of the ANME-1Guaymas lineage. Taxon labels in siena brown indicate thermally ambiguous samples, for example mixed sediments or samples encompassing very steep thermal gradients

enrichments (Holler et al. 2011). These syntrophs form a monophyletic lineage that is distinct from the commonly found sulfate-reducing syntrophs of the *Desulfobacteraceae* and *Desulfobulbaceae* typically dominating mesophilic or cold-seep consortia. The HotSeep-1 bacteria can grow independently in enrichment culture as thermophilic hydrogenotrophic sulfate-reducing bacteria, which enabled their description as "*Candidatus* Desulfofervidus auxilii" (Krukenberg et al. 2016). In symbiotic associations, "*Ca.* Desulfofervidus" forms pili-like cell-to-cell conduits for direct electron exchange with their ANME-1 partners (Wegener et al. 2015). "*Ca.* Desulfofervidus" is not limited to methane-oxidizing consortia, and has been enriched previously in thermophilic butane-oxidizing enrichments from Guaymas Basin (Kniemeyer et al. 2007).

Whereas "*Ca.* Desulfofervidus" occupies the thermophilic niche at 50–60 °C, moderately warm temperatures of 20–37 °C favor an uncultured group within the *Deltaproteobacteria*, the SEEP-SRB2 lineage. These bacteria are frequently found in sediments that are rich in methane and short-chain alkanes, including those of Guaymas Basin (Dowell et al. 2016). The SEEP-SRB2 bacteria have been proposed to function as sulfate-reducing ethane oxidizers (Kleindienst et al. 2012), and indeed occur in Guaymas Basin hydrothermal sediments where short-chain alkanes, especially ethane, are available as substrates (Dowell et al. 2016). Genome sequencing and expression of methane-oxidizing ANME consortia harboring SEEP-SRB2 bacteria have revealed the complete pathway of dissimilatory sulfate reduction, several cytoplasmatic and periplasmatic hydrogenases, and c-type cytochromes that likely function as electron carriers between ANME archaea and their SEEP-SRB2 partners (Krukenberg et al. 2018). So far, a similar role in ethane and alkane oxidation remains to be shown.

Alkane-Oxidizing Bacteria and Archaea in Guaymas Basin

Hydrothermal circulation entrains seawater sulfate into the sediments of Guaymas Basin and makes this electron acceptor available over a wide temperature rage. Therefore, microbial sulfate reduction has adapted to this thermal range and occurs under mesophilic, thermophilic and hyperthermophilic conditions (Jørgensen et al. 1990, 1992; Elsgaard et al. 1994; Weber and Jørgensen 2002; Kallmeyer et al. 2003). Sulfate-reducing bacteria in Guaymas Basin sediments thrive on abundant, low-molecular weight organic substrates, especially acetate (Martens 1990), but aliphatic and aromatic hydrocarbons are also important substrates. For example, studies using cultures of specialized sulfate-reducing bacteria from hydrothermal Guaymas Basin sediments have demonstrated anaerobic short-chain alkane oxidation under mesophilic and thermophilic conditions, i.e. mesophilic sulfate-reducing bacteria that oxidize *n*-butane (Kniemeyer et al. 2007; Jaekel et al. 2013), thermophilic sulfate-reducing cultures oxidizing *n*-butane and propane

(Kniemeyer et al. 2007), and thermophilic sulfate-reducing bacterial isolates oxidizing *n*-decane (Rüter et al. 1994). A detailed overview on hydrocarbon-oxidizing sulfate-reducing bacteria is given in Chap. 2.

Recently, thermophilic butane-oxidizing archaea, termed "*Candidatus* Syntrophoarchaeum spp.", were isolated from Guaymas Basin sediments (Laso-Pérez et al. 2016). These archaea contain a phylogenetically distinct variant of the *mcrA* genes, previously known as the key gene for methane activation in anaerobic methane oxidation, and—in the methanogenic direction—the key gene for reducing the cofactor-bound methyl group and releasing it as methane. This butane-activating *mcrA* gene variant is related to *mcrA* genes found in members of the *Bathyarchaeota* (Evans et al. 2015), however, "*Candidatus* Syntrophoarchaeum" is not a member of the *Bathyarcheota*, but forms a distinct phylogenetic lineage within the *Euryarchaeota* (Laso-Pérez et al. 2016). These euryarchaeal butane oxidizers grow in association with the thermophilic, hydrogenotrophic sulfate-reducing bacterium "*Candidatus* Desulfofervidus auxilii" (Krukenberg et al. 2016). Its syntrophic role is to channel the electrons obtained by butane oxidation to the terminal electron acceptor sulfate. Electron micrographs of "*Ca*. Syntrophoarchaeum" enrichments show a dense network of pili-like nanowires between the cells, strongly suggesting cell-to-cell syntrophic electron transfer by this mechanism (Laso-Pérez et al. 2016, Supplementary materials).

Compared to the sustained efforts that have focused on anaerobic alkane oxidizers in Guaymas Basin, aerobes have received less attention. Aerobic bacterial isolates from Guaymas Basin sediments degraded preferentially aromatic carboxylic acids and to a lesser extent polyaromatic compounds (Bazylinski et al. 1989; Götz and Jannasch 1993); while some of these isolates resembled the aromatics-degrading genus *Cycloclasticus*, they lacked sequence identification. These early studies have recently been followed up by stable isotope probing with polyaromatic substrates, strain enrichment, isolation, and 16S rRNA gene sequencing. Guaymas Basin sediments yielded members of the obligately aromatics-oxidizing genus *Cycloclasticus* and other *Gammaproteobacteria* (*Halomonas*, *Halomonas*, and *Lutibacterium*) that were also frequently recovered from oil seeps and spills (Gutierrez et al. 2015; this topic is further elaborated in Chap. 10).

Genomic Surveys of Guaymas Basin Microbial Communities

Recent metagenomic reconstructions from Guaymas Basin hydrothermal sediments have revealed multiple uncultured bacterial and archaeal phylum- or subphylum-level lineages with the genomic potential for complex carbon substrate and hydrocarbon utilization, fermentative generation of intermediate low-molecular weight compounds, terminal oxidation, carbon fixation, and sulfur cycling

(Dombrowski et al. 2017, 2018). Among the currently ca. 550 reconstructed genomes from Guaymas Basin, new lineages of alkane-oxidizing archaea with unusual *mcrA* genes are promising candidates for novel types of alkane-oxidizing archaea, including ethane oxidizers. The two-pronged approach of comparing alkane-oxidizing enrichments from Guaymas Basin (Wegener et al. 2018) against environmental genomes has great promise for comprehensive genome- and cultivation-based characterizations of these hydrocarbon degraders in Guaymas Basin, and is likely to expand their known phylogenetic and metabolic range.

Guaymas Basin has yielded numerous genomes representing the *Asgardarchaea* (Dombrowski et al. 2018), a new superphylum-level lineage whose phylum-level members are whimsically named for gods of the old Norse pantheon. The *Asgardarchaea* harbor numerous eukaryote-like genes coding proteins for cell architecture, compartmentalization, intracellular vesicle transport and signaling that usually occur in eukaryotes only (Spang et al. 2015; Zaremba-Niedzwiedzka et al. 2017). These unusual archaea appear predisposed as models for an ancestral cell type that engulfed bacterial symbionts and developed the ultrastructural complexity of the eukaryotic cell (Eme et al. 2017). A newly discovered lineage of *Asgardarchaea* from Guaymas Basin, named *Helarchaeota* after the goddess of the underworld, contains the genome complement for mcrA-dependent alkane activation, apparently obtained by lateral gene transfer since other *Asgardarchaea* do not share this genomic blueprint (Seitz et al. 2019). These Guaymas Basin results extend anaerobic alkane-activating pathways widely throughout the archaeal domain.

Challenging Conditions for Microbial Life in Guaymas Basin Hydrothermal Sediments

Our observations and location-specific habitat studies in Guaymas Basin (Biddle et al. 2012; McKay et al. 2012, 2016; Dowell et al. 2016) allow a preliminary synthesis of microbial habitat preferences and activities in this environment, and point to important caveats related to sampling scale and habitat resolution. In the center of hydrothermal hot spots, steep thermal profiles and abundant porewater pools of carbon (DIC, CH_4, LMW organic acids and hydrocarbons), inorganic electron donors (sulfide, potentially hydrogen) and electron acceptors (sulfate) select for microbial populations that can utilize these hydrothermal carbon and energy sources while surviving the combined stresses of thermal extremes, thermal fluctuations, and strongly reducing conditions (Fig. 7).

In the sediment, fluctuating steep geochemical and thermal gradients select for adaptable survival specialists including ANME-1 Guaymas archaea and HotSeep-1 bacteria ("*Candidatus* Desulfofervidus"), whereas on the sediment surface these conditions appear to select for the orange Guaymas *Beggiatoaceae*. On the periphery of hydrothermal hot spots, the thermal and geochemical gradients become smoother, and hydrothermal fluctuations and presumably seawater inmixing are increasingly attenuated. Thermotolerant ANME-1 archaea and HotSeep-1 syntrophs remain

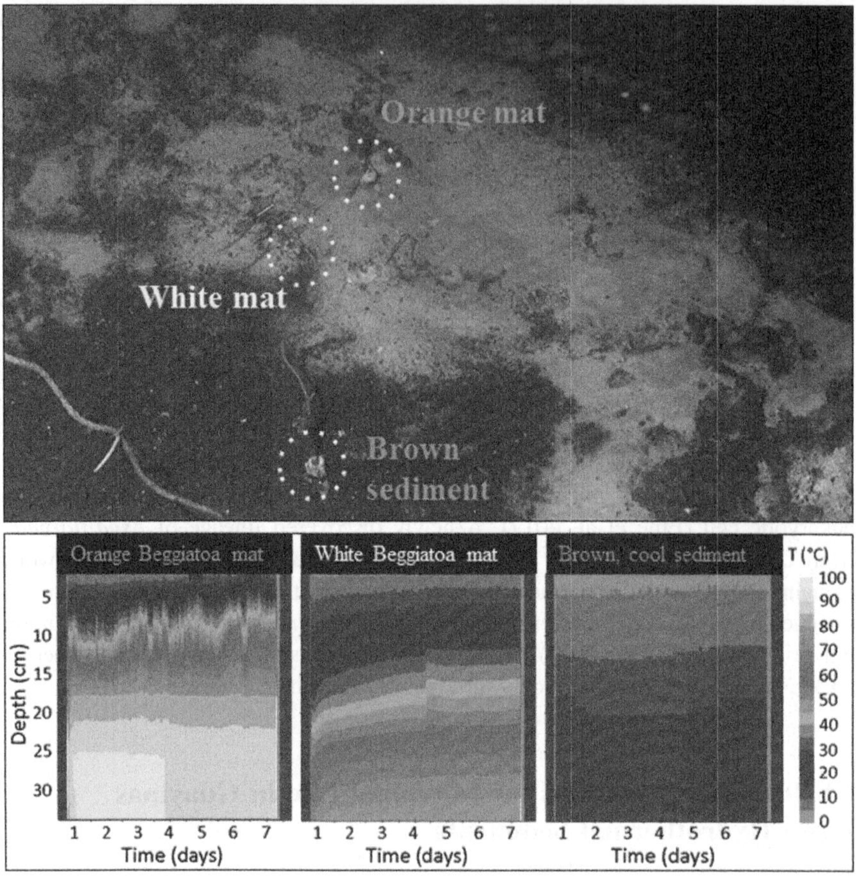

Fig. 7 Thermal regimes across a Guaymas Basin hydrothermal hot spot, recorded by three temperature logging probes inserted into the sediment. The three hydrothermal regimes (central hot spot, periphery, and bare sediment) follow each other from the center of the *Beggiatoaceae* mat dominated by orange filaments to its periphery dominated by white filaments, to the bare sediment surrounding the mat. This sediment may still have some hydrothermal activity, but can no longer sustain a visible microbial mat (McKay et al. 2012). The central mat is characterized by steep geochemical and thermal gradients, and strong thermal fluctuations within 20–60 °C in the upper 10–15 cm sediments (McKay et al. 2016)

detectable, but the ANME-1 Guaymas archaea are found less frequently (McKay et al. 2016). The sediment surface mat changes its composition and is now dominated by large, colorless *Beggiatoaceae* filaments (McKay et al. 2012). Even further towards the outside periphery, hydrothermal temperature and geochemical gradients are strongly attenuated; the visually conspicuous mats of sulfur-oxidizing large filaments disappear, while single-celled sulfur-oxidizing *Epsilonproteobacteria* remain detectable by sequencing, and methane-cycling archaea are gradually replaced by heterotrophic benthic archaea (McKay et al. 2016).

The thermal, chemical, and microbial gradients that change from the center of a hot spot towards its periphery are superimposed on strong downcore gradients, towards increasing temperature and hydrothermal influence at greater sediment depth (Fig. 7). This superimposition of lateral and vertical gradients, together with highly dynamic conditions and hydrothermal circulation in the center of a hot spot, create a complex mosaic of contrasting microhabitats and their inhabitants in the Guaymas Basin sediments. The compact vertical scale, as well as steep biogeochemical and thermal gradients of these microhabitats, as shown by heatflow sensor surveys and microprofiler studies of different mats and sediments in Guaymas Basin (Winkel et al. 2014; Teske et al. 2016), influences the numbers of microbial cells and the amount of DNA and RNA that can be recovered from these sediments. For example, the distribution of cultivable, hyperthermophilic *Thermococcus* and *Pyrococcus* isolates in hydrothermally active Guaymas sediments peaked in the upper sediment layers (Teske et al. 2009). The highest concentration of cultivable cells (10^4 cells/ml) was found in the upper 5 and 10 cm at in situ temperatures up to 75 and 85 °C, respectively. At 13 cm depth, cultivable cells decreased to 10^2 cells/ml, while temperatures increased to ca. 95 °C. Cultivable populations approached the detection limit at 20 cm depth and ca. 105 °C, where an enrichment on undiluted sediment showed initial growth but did not yield any isolates (Teske et al. 2009). Total cell numbers peaked at the sediment surface; cell densities in the upper 5 cm of sediment exceeded 10^9 cells/ml, and quickly decreased by an order of magnitude or more below 5 cm (Meyer et al. 2013). Shipboard measurements of sulfate reduction rates in *Beggiatoaceae*-covered Guaymas Basin sediments showed the highest rates in surface sediments (upper 6 cm) at mesophilic or moderately thermophilic temperatures (Weber and Jørgensen 2002). Further sulfate reduction rate measurements at mesophilic temperatures have consistently detected the highest rates in surficial sediments, mostly the upper 4 cm of hydrothermally active sediments (Meyer et al. 2013; Dowell et al. 2016). Studies based on lipid and DNA recovery have found a similar microbial preference for the surficial sediments at Guaymas Basin. Bacterial and archaeal lipids are most abundant in the surface 5 cm layer (Guezennec et al. 1996; Teske et al. 2002). Archaeal lipid biomarker concentrations decreased by two to three orders of magnitude toward 10–15 cm depth (Schouten et al. 2003). DNA for PCR amplification, cloning, and sequencing of bacterial and archaeal 16S rRNA genes and functional genes for methanogenesis and sulfate reduction (*mcrA*, *dsrAB*) could be extracted from max. 7 cm (*dsrAB*) to 15 cm (*mcrA*) sediment depth (Teske et al. 2002; Dhillon et al. 2003, 2005). These specific depths apply to the sediment cores used in these studies and may change in other sampling locations, but they strongly suggest that the depth of the microbially active surficial sediment layer depends on locally variable hydrothermal fluid and heat flow. Hydrothermal fluctuations flush out biomass and organic carbon from deeper sediments towards the surface layer, in striking contrast to smooth gradients of sedimentary organic matter in non-hydrothermal control sediments (Lin et al. 2017).

Viewing the evidence in context, the highly dynamic hydrothermal sediment habitat of bacteria and archaea in Guaymas Basin is characterized by hydrothermal pulses or high-flux episodes that on occasion purge the microbial community, pushing the survivors towards the sediment surface. By the same token, hydrothermal fluctuations in Guaymas Basin select for microorganisms with a broad thermal range, in individual strains or within a major metabolic group, as evident for methane and hydrocarbon oxidizers. As long as the hydrothermal sediments of Guaymas Basin provide sufficient carbon and energy sources to sustain high-temperature microbial growth, and prevent supply bottlenecks that quickly become fatal at high temperatures (Lloyd et al. 2005), thermophilic or at least heat-tolerant variants of classic cold-seep microbial populations will thrive in this energy-rich habitat, under the condition that they are sufficiently resilient to hydrothermal purges sweeping through the sediments. Increased temperatures and microbial process rates linked to faster growth capabilities may explain why the hydrothermal sediments of Guaymas Basin, regardless of difficult access, have become a particularly successful field site for the enrichment and isolation of demanding, usually slow-growing anaerobic hydrocarbon degraders (Teske 2019; Wegener et al. 2018). The hydrothermal heat that sets Guaymas Basin apart from cold seep sites selects for microbial growth over a wide temperature range, and increases metabolic rates—both desirable traits for cultivations and potential applications in the future.

References

Bazylinski DA, Farrington JW, Jannasch HW (1988) Hydrocarbons in surface sediments from a Guaymas Basin hydrothermal vent site. Org Geochem 12:547–558

Bazylinski DA, Wirsen CO, Jannasch HW (1989) Microbial utilization of naturally occurring hydrocarbons at the Guaymas Basin hydrothermal vent site. Appl Environ Microbiol 55:2832–2836

Berndt C, Hensen C, Mortera-Gutierrez C, Sarkar S, Geilert S, Schmidt M, Liebetrau V, Kipfer R, Scholz F, Doll M, Muff S, Karstens J, Planke S, Petersen S, Böttner C, Chi WC, Moser M, Behrendt R, Fiskal A, Lever MA, Su CC, Deng L, Brennwald MS, Lizarralde D (2016) Rifting under steam—how magmatism triggers methane venting from sedimentary basins. Geology 44:767–770

Biddle JF, Cardman Z, Mendlovitz H, Albert DB, Lloyd KG, Boetius A, Teske A (2012) Anaerobic oxidation of methane at different temperature regimes in Guaymas Basin hydrothermal sediments. ISME J 6:1018–1031

Bowles MW, Nigro LM, Teske AP, Joye SB (2012) Denitrification and environmental factors influencing nitrate removal in Guaymas Basin hydrothermally-altered sediments. Front Microbiol 3:377

Buckley A, MacGregor BJ, Teske A (2019) Identification, expression and activity of candidate nitrite reductases from orange *Beggiatoaceae*. Guaymas Basin. Front Microbiol 10:644

Bult CJ, White O, Olsen GJ, Zhou L, Fleischmann RD, Sutton GG, Blake JA, FitzGerald LM, Clayton RA, Gocayne JD, Kerlavage AR, Dougherty BA, Tomb J-F, Adams MD, Reich CI, Overbeek R, Kirkness EF, Weinstock KG, Merrick JM, Glodek A, Scott JL, Geoghagen NSM, Weidman JF, Fuhrmann JL, Nguyen D, Utterback TR, Kelley JM, Peterson JD, Sadow PW, Hanna MC, Cotton MD, Roberts KM, Hurst MA, Kaine BP, Borodovsky M, Klenk H-P, Fraser CM, Smith HO, Woese CR, Venter JC (1996) Complete genome sequence of the methanogenic archaeon, *Methanococcus jannaschii*. Science 272:1058–1073

Calvert SE (1964) Factors affecting distribution of laminated diatomaceous sediments in the Gulf of California. In: Van Andel TH, Shor GG Jr (eds) Marine geology of the Gulf of California, vol 3. American Association of Petroleum Geologists Memoir, Tulsa, pp 311–330

Calvert SE (1966) Origin of diatom-rich, varved sediments from the Gulf of California. J Geol 74:546–565

Campbell BJ, Engel AS, Porter ML, Takai K (2006) The versatile *epsilon proteobacteria*: key players in sulphidic habitats. Nat Rev Microbiol 4:458–468

Curray JR, Moore DG, Aguayo EJ (1979) Leg 64 seeks evidence on development of basins. Geotimes 24:18–20

Curray JR, Moore DG (1982) Initial reports of the Deep Sea Drilling Project, vol 64. U.S. Government Printing Office, Washington, DC. https://doi.org/10.2973/dsdp.proc.64.1982

Dekas AE, Poretsky RS, Orphan VJ (2009) Deep-sea archaea fix and share nitrogen in methane-consuming microbial consortia. Science 326:422–426

De la Lanza-Espino G, Soto LA (1999) Sedimentary geochemistry of hydrothermal vents in Guaymas Basin, Gulf of California, Mexico. Appl Geochem 14:499–510

Dhillon A, Teske A, Dillon J, Stahl DA, Sogin ML (2003) Molecular characterization of sulfate-reducing bacteria in the Guaymas Basin. Appl Environ Microbiol 69:2765–2772

Dhillon A, Lever M, Lloyd KG, Albert DB, Sogin ML, Teske A (2005) Methanogen diversity evidenced by molecular characterization of methyl coenzyme M reductase A genes (*mcr*A) in hydrothermal sediments of the Guaymas Basin. Appl Environ Microbiol 71:4592–4601

Didyk BM, Simoneit BRT (1989) Hydrothermal oil of Guaymas Basin and implications for petroleum formation mechanisms. Nature 342:65–69

Dombrowski N, Seitz K, Teske A, Baker BJ (2017) Genomic insights into potential interdependencies in microbial hydrocarbon and nutrient cycling in hydrothermal sediment communities. Microbiome 5:106

Dombrowski N, Teske AP, Baker BJ (2018) Extensive microbial metabolic diversity and redundancy in Guaymas Basin hydrothermal sediments. Nat Commun 9:4999

Dowell F, Cardman Z, Dasarathy S, Kellermann MY, Lipp JS, Ruff SE, Biddle JF, Lloyd KG, Hinrichs K-U, Albert DB, Mendlovitz H, Teske A (2016) Microbial communities in methane- and short chain alkane-rich hydrothermal sediments of Guaymas Basin. Front Microbiol 7:17

Einsele G, Gieskes JM, Curray J, Moore DM, Aguayo E, Aubry M-P, Fornari D, Guerrero J, Kastner M, Kelts K, Lyle M, Matoba Y, Molina-Criz A, Niemitz J, Rueda J, Saunders A, Schrader H, Simoneit B, Vacquier V (1980) Intrusion of basaltic sills into highly porous sediments and resulting hydrothermal activity. Nature 283:441–445

Elsgaard L, Isaksen MF, Jørgensen BB, Alayse A-M, Jannasch HW (1994) Microbial sulfate reduction in deep-sea sediments at the Guaymas Basin hydrothermal vent area: influence of temperature and substrates. Geochim Cosmochim Acta 58:3335–3343

Eme L, Spang A, Lombard J, Stairs CW, Ettema TJG (2017) Archaea and the origin of eukaryotes. Nat Rev Microbiol 15:711–723

Evans PN, Parks DH, Chadwick GL, Robbins SJ, Orphan VJ, Golding SD, Tyson GW (2015) Methane metabolism in the archaeal phylum Bathyarchaeota revealed by genome-centric metagenomics. Science 350:434–438

Fisher AT, Becker K (1991) Heat flow, hydrothermal circulation and basalt intrusions in the Guaymas Basin, Gulf of California. Earth Plan Sci Lett 103:84–89

Galimov EM, Simoneit BRT (1982) Geochemistry of interstitial gases in sedimentary deposits of the Gulf of California, Deep Sea Drilling Project Leg 64. In: Curray JR, Blakeslee J, Platt LW, Stout LN, Moore DG, Aguayo JE et al (eds) Initial reports of the Deep Sea Drilling Project, vol 64. U.S. Government Printing Office, Washington, DC, pp 781–787

Gieskes JM, Kastner M, Einsele G, Kelts K, Niemitz J (1982) Hydrothermal activity in the Guaymas Basin, Gulf of California: a synthesis. In Curray JR, Blakeslee J, Platt LW, Stout LN, Moore DG, Aguayo JE et al (eds) Initial reports of the Deep Sea Drilling Project, vol 64. U.S. Government Printing Office, Washington, DC, pp 1159–1167

Götz FE, Jannasch HW (1993) Aromatic hydrocarbon-degrading bacteria in the petroleum-rich sediments of Guaymas Basin hydrothermal vent site: preference for aromatic carboxylic acids. Geomicrobiol J 11:1–18

Guezennec JG, Dussauze J, Bian M, Rocchicioli F, Ringelberg D, Hedrick DB, White DC (1996) Bacterial community structure from Guaymas Basin, Gulf of California, as determined by analysis of phospholipid ester-linked fatty acids. J Mar Biotechnol 4:165–175

Gundersen JK, Jørgensen BB, Larsen E, Jannasch HW (1992) Mats of giant sulphur bacteria on deep-sea sediments due to fluctuating hydrothermal flow. Nature 360:454–456

Gutierrez T, Biddle JF, Teske A, Aitken MD (2015) Cultivation-dependent and cultivation-independent characterization of hydrocarbon-degrading bacteria in Guaymas Basin sediments. Front Microbiol 6:695

He Y, Xiao X, Wang F (2013) Metagenome reveals potential microbial degradation of hydrocarbon coupled with sulfate reduction in an oil-immersed chimney from Guaymas Basin. Front Microbiol 4:148

Holler T, Widdel F, Knittel K, Amann R, Kellermann MY, Hinrichs K-U, Teske A, Boetius A (2011) Thermophilic anaerobic oxidation of methane by marine microbial consortia. ISME J 5:1946–1956

Jaekel U, Musat N, Adam B, Kuypers M, Grundmann O, Musat F (2013) Anaerobic degradation of propane and butane by sulfate-reducing bacteria enriched from marine hydrocarbon cold seeps. ISME J 7:885–895

Jannasch HW, Nelson DC, Wirsen CO (1989) Massive natural occurrence of unusually large bacteria (*Beggiatoa* spp.) at a hydrothermal deep-sea vent site. Nature 342:834–836

Jeanthon C, L'Haridon S, Pradel N, Prieur D (1999) Rapid identification of hyperthermophilic methanococci isolated from deep-sea hydrothermal vents. Int J Syst Bacteriol 49:591–594

Jones WJ, Leigh JA, Mayer F, Woese CR, Wolfe RS (1983) *Methanococus jannaschii* sp. nov., an extremely thermophilic methanogen from a submarine hydrothermal vent. Arch Microbiol 136:254–261

Jones WJ, Stugard CE, Jannasch HW (1989) Comparison of thermophilic methanogens from submarine hydrothermal vents. Arch Microbiol 151:314–319

Jørgensen BB, Isaksen MF, Jannasch HW (1992) Bacterial sulfate reduction above 100 °C in deep-sea hydrothermal vent systems. Science 258:1756–1757

Jørgensen BB, Zawacki LX, Jannasch HW (1990) Thermophilic bacterial sulfate reduction in deep-sea sediments at the Guaymas Basin hydrothermal vents (Gulf of California). Deep Sea Res I 37:695–710

Kallmeyer J, Ferdelman TG, Jansen K-H, Jørgensen BB (2003) A high pressure thermal gradient block for investigating microbial activity in multiple deep-sea samples. J Microbiol Methods 55:165–172

Kallmeyer J, Boetius A (2004) Effects of temperature and pressure on sulfate reduction and anaerobic oxidation of methane in hydrothermal sediments of Guaymas Basin. Appl Environ Microbiol 70:1231–1233

Kastner M (1982) Evidence for two distinct hydrothermal systems in the Guaymas Basin. In: Curray JR, Blakeslee J, Platt LW, Stout LN, Moore DG, Aguayo JE, et al (eds) Initial reports of the Deep Sea Drilling Project, vol 64. U.S. Government Printing Office, Washington, DC, pp 1143–1158

Kawka OE, Simoneit BRT (1987) Survey of hydrothermally generated petroleums from the Guaymas Basin spreading center. Org Geochem 11:311–328

Kellermann MY, Wegener G, Elvert M, Yoshinaga MY, Lin YS, Holler T, Mollar XP, Knittel K, Hinrichs K-U (2012) Autotrophy as predominant mode of carbon fixation in thermophilic anaerobic methane-oxidizing microbial communities. Proc Natl Acad Sci U S A 109:19321–19326

Kleindienst S, Ramette A, Amann R, Knittel K (2012) Distribution and in situ abundance of sulfate-reducing bacteria in diverse marine hydrocarbon seep sediments. Environ Microbiol 14:2689–2710

Kniemeyer O, Musat F, Sievert SM, Knittel K, Wilkes H, Blumenberg M, Michaelis W, Classen A, Bolm C, Joye SB, Widdel F (2007) Anaerobic oxidation of short-chain hydrocarbons by marine sulphate-reducing bacteria. Nature 449:898–901

Knittel K, Boetius A (2009) Anaerobic oxidation of methane: progress with an unknown process. Annual Rev Microbiol 63:311–334

Krukenberg V, Harding K, Richter M, Glockner FO, Gruber-Vodicka HR, Adam B, Berg JS, Knittel K, Tegetmeyer HE, Boetius A, Wegener G (2016) *Candidatus* Desulfofervidus auxilii, a hydrogenotrophic sulfate-reducing bacterium involved in the thermophilic anaerobic oxidation of methane. Environ Microbiol 18:3073–3091

Krukenberg V, Riedel D, Gruber-Vodicka HR, Buttigieg PL, Tegetmeyer HE, Boetius A, Wegener G (2018) Gene expression and ultrastructure of meso- and thermophilic methanotrophic consortia. Environ Microbiol 20:1651–1666

Kurr M, Huber R, König H, Jannasch HW, Fricke H, Trincone A et al (1991) *Methanopyrus kandleri*, gen. and sp. nov. represents a novel group of hyperthermophilic methanogens, growing at 110 °C. Arch Microbiol 156:239–247

Laso-Pérez R, Wegener G, Knittel K, Widdel F, Harding KJ, Krukenberg V, Meier DV, Richter M, Tegetmeyer HE, Riedel D, Richnow H-H, Adrian L, Reemtsma T, Lechtenfeld O, Musat F (2016) Thermophilic archaea activate butane via alkyl-coenzyme M formation. Nature 539:396–401

Lever MA, Teske A (2015) Methane-cycling archaeal diversity in hydrothermal sediment investigated by general and group-specific functional gene and 16S rRNA gene PCR primers. Appl Environ Microbiol 81:1426–1441

Lin Y-S, Koch BP, Feseker T, Ziervogel K, Goldhammer T, Schmidt S, Witt M, Kellermann M, Zabel M, Teske A, Hinrichs K-U (2017) Near-surface heating of young rift sediment causes mass production and discharge of reactive dissolved organic matter. Sci Rep 7:44864

Lizarralde D, Soule SA, Seewald J, Proskurowski G (2011) Carbon release by off-axis magmatism in a young sedimented spreading centre. Nat Geosci 4:50–54

Lloyd KG, Edgcomb VP, Molyneaux SJ, Boer S, Wirsen CO, Atkins MS, Teske A (2005) Effect of dissolved sulfide, pH, and temperature on growth and survival of marine hyperthermophilic archaea. Appl Environ Microbiol 71:6383–6387

Lonsdale P, Becker K (1985) Hydrothermal plumes, hot springs, and conductive heat flow in the Southern Trough of Guaymas Basin. Earth Planet Sci Lett 73:211–225

Luther GW, Rozan TF, Taillefert M, Nuzzio DB, Di Meo C, Shank TM, Lutz RA, Cary SC (2001) Chemical speciation drives hydrothermal vent ecology. Nature 410:813–816

MacGregor BJ, Biddle JF, Siebert JR, Staunton E, Hegg E, Matthysse AG, Teske A (2013a) Why orange Guaymas Basin *Beggiatoa* spp. are orange: single- filament genome-enabled identification of an abundant octaheme cytochrome with hydroxylamine oxidase, hydrazine oxidase and nitrite reductase activities. Appl Environ Microbiol 79:1183–1190

MacGregor BJ, Biddle JF, Harbort C, Matthysse AG, Teske A (2013b) Sulfide oxidation, nitrate respiration, carbon acquisition and electron transport pathways suggested by the draft genome of a single orange Guaymas Basin Beggiatoa (Cand. Maribeggiatoa) sp. filament. Mar Genomics 11:53–65

MacGregor BJ, Biddle JF, Teske A (2013c) Mobile elements in a single-filament orange Guaymas Basin *Beggiatoa* ("Candidatus Maribeggiatoa") sp. draft genome; evidence for genetic exchange with cyanobacteria. Appl Environ Microbiol 79:3974–3985

Martens CS (1990) Generation of short chain organic acid anions in hydrothermally altered sediments of the Guaymas Basin, Gulf of California. Appl Geochem 5:71–76

McGlynn SE, Chadwick GL, Kempes CP, Orphan VJ (2015) Single cell activity reveals direct electron transfer in methanotrophic consortia. Nature 526:531–535

McKay LJ, MacGregor BJ, Biddle JF, Mendlovitz HP, Hoer D, Lipp JS, Lloyd KG, Teske AP (2012) Spatial heterogeneity and underlying geochemistry of phylogenetically diverse orange and white *Beggiatoa* mats in Guaymas Basin hydrothermal sediments. Deep Sea Res I 67:21–31

McKay L, Klokman VW, Mendlovitz HP, LaRowe DE, Hoer DR, Albert D, de Beer J, Amend J, Teske A (2016) Thermal and geochemical influences on microbial biogeography in the hydrothermal sediments of Guaymas Basin. Environ Microbiol Rep 8:150–161

McKay LJ, Hatzenpichler R, Inskeep WP, Fields MW (2017) Occurrence and expression of novel methyl-coenzyme M reductase gene (mcrA) variants in hot spring sediments. Scientific Reports 7:725

Meyer S, Wegener G, Lloyd KG, Teske A, Boetius A, Ramette A (2013) Microbial habitat connectivity across spatial scales and hydrothermal temperature gradients at Guaymas Basin. Front Microbiol 4:207

Miyazaki J, Higa R, Toki T, Ashi J, Tsunogai U, Nunoura T, Imachi H, Takai K (2009) Molecular characterization of potential nitrogen fixation by anaerobic methane-oxidizing archaea in the methane seep sediments at the Number 8 Kumano Knoll in the Kumano Basin, offshore of Japan. Appl Environ Microbiol 75:7153–7162

Møller MM, Nielsen LP, Jørgensen BB (1985) Oxygen responses and mat formation of *Beggiatoa* spp. Appl Environ Microbiol 50:373–382

Mulder A, van de Graaf AA, Robertson LA, Kuenen JG (1995) Anaerobic ammonium oxidation discovered in a denitrifying fluidized bed reactor. FEMS Microbiol Ecol 16:177–183

Nelson DC, Castenholz RW (1981) Use of reduced sulfur compounds by *Beggiatoa* sp. J Bacteriol 147:140–154

Nelson DC, Revsbech NP, Jørgensen BB (1986) Microoxic-anoxic niche of *Beggiatoa* spp.: microelectrode survey of marine and freshwater strains. Appl Environ Microbiol 52:161–168

Nelson DC, Wirsen CO, Jannasch HW (1989) Characterization of large, autotrophic *Beggiatoa* spp. abundant at hydrothermal vents of the Guaymas Basin. Appl Environ Microbiol 55:2909–2917

Nobu MK, Narihiro T, Kuroda K, Mei R, Liu WT (2016) Chasing the elusive Euryarchaeota class WSA2: Genomes reveal a uniquely fastidious methyl-reducing methanogen. ISME J 10:2478–2487

Oremland RS, Culbertson C, Simoneit BRT (1982) Methanogenic activity in sediment from leg 64, Gulf of California. In: Curray JR, Blakeslee J, Platt LW, Stout LN, Moore DG, Aguayo JE et al (eds) Initial reports of the Deep Sea Drilling Project, vol 64. U.S. Government Printing Office, Washington, DC, pp 759–762

Pagé A, Tivey KK, Stakes DS, Reysenbach A-L (2008) Temporal and spatial archaeal colonization of hydrothermal deposits. Environ Microbiol 10:874–884

Pearson A, Seewald JS, Eglinton TI (2005) Bacterial incorporation of relict carbon in the hydrothermal environment of Guaymas Basin. Geochim Cosmochim Acta 69:5477–5486

Pernthaler A, Dekas AE, Brown CT, Goffredi SK, Embaye T, Orphan VJ (2008) Diverse syntrophic partnerships from deep-sea methane vents revealed by direct cell capture and metagenomics. Proc Natl Acad Sci U S A 105:7052–7057

Peter JM, Peltonen P, Scott SD, Simoneit BRT, Kawka OE (1991) ^{14}C ages of hydrothermal petroleum and carbonate in Guaymas Basin, Gulf of California: implications for oil generation, expulsion, and migration. Geology 19:253–256

Rüter P, Rabus R, Wilkes H, Aeckersberg F, Rainey FA, Jannasch HW, Widdel F (1994) Anaerobic oxidation of hydrocarbons in crude oil by new types of sulphate-reducing bacteria. Nature 372:455–458

Rullkötter J, von der Dick H, Welte DH (1982) Organic petrography and extractable hydrocarbons of sediment from the Gulf of California, Deep Sea Drilling Project Leg 64. In: Curray JR, Blakeslee J, Platt LW, Stout LN, Moore DG, Aguayo JE et al (eds) Initial reports of the Deep Sea Drilling Project, vol 64. U.S. Government Printing Office, Washington, DC, pp 837–853

Russ L, Kartal B, op den Camp HJM, Sollai M, Le Bruchec J, Caprais J-C, Godfroy A, Sinninghe Damsté JS, Jetten MSM (2013) Presence and diversity of anammox bacteria in cold hydrocarbon-rich seeps and hydrothermal vent sediments of the Guaymas Basin. Front Microbiol 4:219

Saunders A, Fornari DJ, Joron JL, Tarney J, Treuil M (1982) Geochemistry of basic igneous rocks, Gulf of California. In: Curray JR, Blakeslee J, Platt LW, Stout LN, Moore DG, Aguayo JE et al (eds) Initial reports of the Deep Sea Drilling Project, vol 64. U.S. Government Printing Office, Washington, DC, pp 595–642

Schouten S, Wakeham SG, Hopmans EC, Sinninghe Damste JS (2003) Biogeochemical evidence that thermophilic archaea mediate the anaerobic oxidation of methane. Appl Environ Microbiol 69:1680–1686

Schutte C, Teske A, MacGregor BJ, Salman-Carvalho V, Lavik G, Hach P, de Beer D (2018) Filamentous giant *Beggiatoaceae* from Guaymas Basin are capable of both denitrification and dissimilatory nitrate reduction to ammonium (DNRA). Appl Environ Microbiol 84:e02860–17

Seitz KW, Dombrowski N, Eme L, Spang A, Lombard J, Sieber J, Teske AP, Ettema TJG, Baker BJ (2019) Asgard Archaea capable of anaerobic hydrocarbon cycling. Nat Commun 10:1822

Simoneit BRT, Bode GR (1982) Appendix II: carbon/carbonate and nitrogen analysis, Leg 64, Gulf of California. In: Curray JR, Blakeslee J, Platt LW, Stout LN, Moore DG, Aguayo JE et al (eds) Initial reports of the Deep Sea Drilling Project, vol 64. U.S. Government Printing Office, Washington, DC, pp 1303–1305

Simoneit BRT, Lonsdale PF (1982) Hydrothermal petroleum in mineralized mounds at the seabed of Guaymas Basin. Nature 295:198–202

Spang A, Saw JH, Jørgensen SL, Zaremba-Niedzwiedzka K, Martijn J, Lind AE, van Eijk R, Schleper C, Guy L, Ettema TJG (2015) Complex archaea that bridge the gap between prokaryotes and eukaryotes. Nature 521:173–179

Takai K, Nakamura K, Tori T, Tsunogai U, Miyazaki M, Miyazaki J, Hirayama H, Nakagawa S, Nunoura T, Horikoshi K (2008) Cell proliferation at 122 °C and isotopically heavy CH_4 production by a hyperthermophilic methanogen under high- pressure cultivation. Proc Natl Acad Sci USA 105:10949–10954

Teske A (2019) Hydrocarbon-degrading microbial communities in natural oil seeps. In: McGenity TJ (ed) Handbook of hydrocarbon and lipid microbiology. Microbial communities utilizing hydrocarbons and lipids: members, metagenomic and ecophysiology. Springer

Teske A, Salman V (2014) The family *Beggiatoaceae.* In: Rosenberg E, DeLong EF, Thompson F, Lory S, Stackebrandt E (eds) The Prokaryotes—Gammaproteobacteria. The Prokaryotes, 4th edn. Springer, Berlin/Heidelberg, pp 93–134

Teske A, Lizarralde D, Höfig TW (2018) Expedition 385 scientific prospectus: Guaymas Basin tectonics and biosphere. Int Ocean Discov Program. https://doi.org/10.14379/iodp.sp.385.2018

Teske A, Hinrichs K-U, Edgcomb VP, de Vera Gomez A, Kysela D, Sylva SP, Sogin ML, Jannasch HW (2002) Microbial diversity in hydrothermal sediments in the Guaymas Basin: evidence for anaerobic methanotrophic communities. Appl Environ Microbiol 68:1994–2007

Teske A, Edgcomb V, Rivers AR, Thompson JR, de Vera Gomez A, Molyneaux SJ, Wirsen CO (2009) A molecular and physiological survey of a diverse collection of hydrothermal vent *Thermococcus* and *Pyrococcus* isolates. Extremophiles 13:917–923

Teske A, Callaghan AV, LaRowe D (2014) Biosphere frontiers of subsurface life in the sedimented hydrothermal system of Guaymas Basin. Front Microbiol 5:362

Teske A, de Beer D, McKay L, Tivey MK, Biddle JF, Hoer D, Lloyd KG, Lever MA, Røy H, Albert DB, Mendlovitz H, MacGregor BJ (2016) The Guaymas Basin hiking guide to hydrothermal mounds, chimneys and microbial mats: complex seafloor expressions of subsurface hydrothermal circulation. Front Microbiol 7:75

Teske A, McKay LJ, Ravelo AC, Aiello I, Mortera C, Núñez-Useche F, Canet C, Chanton J, Brunner B, Hensen C, Ramirez GA, Sibert RJ, Turner T, White D, Chambers CR, Buckley A, Joye SB, Soule SA, Lizarralde D (2019) Characteristics and evolution of sill-driven off-axis hydrothermalism in Guaymas Basin – the Ringvent site. Sci Rep 9:13847

Von Damm KL, Edmond JM, Measures CI, Grant B (1985) Chemistry of submarine hydrothermal solutions at Guaymas Basin, Gulf of California. Geochim Cosmochim Acta 49:2221–2237

Weber A, Jørgensen BB (2002) Bacterial sulfate reduction in hydrothermal sediments of the Guaymas Basin, Gulf of California, Mexico. Deep Sea Res I 149:827–841

Wegener G, Krukenberg V, Riedel D, Tegetmeyer HE, Boetius A (2015) Intracellular wiring enables electron transfer between methanotrophic archaea and bacteria. Nature 526:587–590

Wegener G, Laso-Pérez R, Hahn C, Liebeke M, Teske A, Knittel K (2018) Anaerobic enrichment cultures from the Guaymas Basin reveal an unexpected diversity of thermophilic hydrocarbon-oxidizing archaea. In: Abstract at ISME17, International Symposium on Microbial Ecology, Leipzig, Germany, 12–17 Aug 2018

Welhan JA (1988) Origins of methane in hydrothermal systems. Chem Geol 71:183–198

Winkel M, De Beer D, Lavik G, Peplies J, Mussmann M (2014) Close association of active nitrifyers with *Beggiatoa* mats covering deep-sea hydrothermal sediments. Environ Microbiol 16:1612–1626

Zaremba-Niedzwiedzka K, Caceres EF, Saw JH, Bäckström D, Juzokaite L, Vancaester E, Seitz KW, Anantharaman K, Starnawski P, Kjeldsen KU, Stott MB, Nunoura T, Banfield JF, Schramm A, Baker BJ, Spang A, Ettema TJG (2017) Asgard archaea illuminate the origin of eukaryotic cellular complexity. Nature 541:353–358

Chapter 4
The Gulf of Mexico: An Introductory Survey of a Seep-Dominated Seafloor Landscape

Andreas Teske and Samantha B. Joye

Abstract The Gulf of Mexico is home to numerous hydrocarbon seeps, brine lakes and mud volcanoes that provide habitats for hydrocarbon-dependent microbial communities; trophic linkages connect these hydrocarbon microbiota with benthic marine invertebrates and provide the foundation of complex benthic ecosystems that are ultimately sustained by hydrocarbon seepage. Sampling and reconnaissance by submersibles provide a first impression of the diversity of Gulf of Mexico hydrocarbon seeps, paving the way for further discoveries. This chapter provides on overview of selected hydrocarbon and brine seeps on the northern slope of the Gulf of Mexico that were visited and explored during two month-long cruises with research submersible *Alvin* and R/V *Atlantis* in 2010 and 2014.

Bathymetry and Fluid Geochemistry in the Gulf of Mexico

The sediments of the continental slope of the northern Gulf of Mexico contain large reservoirs of petroleum and gas that sustain an arc of seafloor hydrocarbon seeps on the continental slope from Mississippi to Texas. The Gulf of Mexico seeps include many classic examples and study sites for cold seep communities, as introduced in Chap. 1; they also include highly unusual habitats that not easily found elsewhere, such as the asphalt volcanoes discussed in Chap. 5. In contrast to the destruction wrought by the Deepwater Horizon oil well blowout, natural seeps in the Gulf of Mexico gradually discharge hydrocarbons and these substrate fuel dynamic and complex seafloor ecosystems with conspicuous communities of free-living microbes that often form mats, as well as with microbial symbionts that sustain

A. Teske (✉)
University of North Carolina at Chapel Hill, Chapel Hill, NC 27599-3300, USA
e-mail: teske@email.unc.edu

S. B. Joye
University of Georgia, Athens, GA 30602-3636, USA
e-mail: mjoye@uga.edu

A. Teske and V. Carvalho (eds.), *Marine Hydrocarbon Seeps*,
Springer Oceanography, https://doi.org/10.1007/978-3-030-34827-4_4

dense accumulations of invertebrate hosts McDonald et al. (1990). These seafloor environments are characterized by petroleum leakage and channelized gas flux and ebullition from the seafloor (Kennicutt et al. 1988; Brooks et al. 1986), by gas hydrate formation driven by methane seepage (Brooks et al. 1984; Sassen et al. 1999), by brine and fluidized mud discharge (Roberts and Carney 1997), and/or by methane-derived authigenic carbonates as the principal hard substrate along the seabed (Roberts et al. 2010a). Fluid and gas flow from subsurface hydrocarbon reservoirs can locally entrain and mobilize sediment, resulting in hydrocarbon-rich mud volcanoes on the seafloor (Roberts and Carney 1997; McDonald et al. 2000).

The northern slope of the Gulf of Mexico harbors numerous seep sites strongly influenced by subseafloor brines. The Mid-Jurassic Louann evaporite formation, predominantly halite, extends from Texas to the Florida panhandle underneath the northern Gulf slope, coast and coastal plain (Amos 1987). These salt formations have a lower density than the overlying compacted marine sediments (Lerche and Petersen 1995). Since they are incompressible, they gradually move and buckle upwards through the surrounding and overlying sediment layers, and create structurally complex salt dome crests and mounds in the overlying seafloor (Roberts et al. 1990) that result in a highly conspicuous and characteristic, topsy-turvy seafloor bathymetry (Bryant et al. 1990). This seafloor landscape of domes, basins, steep ridges and slopes extends from the shelf break downslope towards the deep basin of the Gulf of Mexico (Fig. 1). It is dissected by fractures within the thick sediment layers and salt deposits that open up pathways for upward migration of subsurface brines and hydrocarbons in equal measure (Roberts et al. 1999).

Depending on the dynamics of their upward movement, hydrocarbon-rich brines collect gradually in seafloor depressions as quiescent, anoxic brine lakes, or they

Fig. 1 Northern Gulf of Mexico slope bathymetry. The map is publicly available from the Bureau of Ocean Energy Management (BOEM) at www.boem.gov/Gulf-of-Mexico-Bathymetry [retrieved on August 8, 2017]. The bathymetry grid was created from 3D seismic surveys; it defines water depth with 1.4 billion 40-by-40 ft cells and is available in feet and meters. BOEM grid coverage is the area defined by the color in this image. Shaded relief is vertically exaggerated by a factor of five. The map was modified to show the positions of seep sites discussed in this chapter

emerge as highly dynamic mud volcanoes. Whereas brine lakes have a stable halocline and redoxcline at the brine/seawater interface that favors colonization by chemosynthetic invertebrate communities, for example mussels forming extensive banks along the lake shore (MacDonald et al. 1990), mud volcanoes are pulsating with fluidized mud that frequently overflows their seafloor basin and is running downslope over the surrounding seafloor in a network of briny rivulets; these conditions do not favor colonization by invertebrates (McDonald et al. 2000; Joye et al. 2005).

Brine lakes and mud volcanoes are not mutually exclusive forms of hydrocarbon seeps but cycle through phases; a currently quiescent brine lake can have an eventful past as an active mud volcano (MacDonald and Peccini 2009), and may erupt again after extended quiescence, or in irregular intervals (McDonald et al. 2000). Given their physical linkage to subseafloor salt deposits, briny seafloor lakes and mud volcanoes provide extensive and diverse habitats for halophiles and extreme halophiles, and select for halophilic or at least halotolerant variants of hydrocarbon-utilizing or chemosynthetic bacteria and archaea (Lloyd et al. 2006; Joye et al. 2009). As conduits of fluid migration from the subsurface to the seafloor, mud volcanoes also provide windows into the deep subseafloor biosphere—this aspect of their microbiology remains to be explored further (Hoshino et al. 2017).

Microbial Communities and Their Environmental Constraints in Gulf of Mexico Sediments

The chemistry of the discharging fluids varies within and between sites. Different combinations of brine, gaseous and liquid hydrocarbons, sulfides, dissolved inorganic carbon, nutrients and low-molecular weight organic acids select for specifically-adapted microbial communities dominated by specialized taxa. The microbial signature of seeps contrasts with non-seep habitats that are sustained by metabolism of sedimentary organic matter of photosynthetic origin (Lloyd et al. 2010). Seep-adapted microbial communities include anaerobic methane-oxidizing archaea (Lloyd et al. 2006) and hydrocarbon-degrading sulfate-reducing bacteria (Lloyd et al. 2010) within the sediment, and extensive chemosynthetic sulfur-oxidizing microbial mats occur at the sediment-water interface (Sassen et al. 1993; Nikolaus et al. 2003). The hydrocarbon-rich Gulf of Mexico seeps have yielded enrichment cultures of bacteria and archaea capable of oxidizing light alkanes up to butane (Kniemeyer et al. 2007); Chap. 2 provides a global overview on the wide range of anaerobic, alkane-oxidizing organisms. The first known anaerobic ethane-oxidizing archaeum, "*Candidatus* Argoarchaeum ethanivorans", was isolated from Gulf of Mexico seep sediments (Chen et al. 2019). In syntrophic association with sulfate-reducing bacteria, "*Ca.* Argoarchaeum" oxidizes ethane to CO_2 using a modified version of methyl coenzyme reductase to activate ethane to

Table 1 Selected porewater geochemical parameters in seawater versus brines from the Gulf of Mexico

Parameter	Salinity	Cl^-	$SO_4^=$	HS^-	NH_4^+	NO_x	PO_4^{3-}	CH_4	DOC
Seawater	35	559	28	0	<1	<5	<1	0.004	25–50
Halite-derived	90–200	>750	0	<0.1	>2000	<50	~5	>150	>1000
Hydrate-derived	90–200	>750	5–50	>0.5	<100	<10	<5	>150	<1000

Data from Joye et al. (2005, 2010), Bowles et al. (2016), and Joye et al. (unpublished)

ethyl-coenzyme M (Chen et al. 2019), analogous to activation of butane and propane by archaea in the phylogenetically distinct "*Ca.* Syntrophoarchaeum" lineage (Laso-Pérez et al. 2016).

One of the most significant stressors for seep communities on the northern Gulf of Mexico slope is exposure to the high salinity of subseafloor brines (Table 1). Relative to seawater, brines are quite distinct. Seafloor brines can have a very high salinity—up to 200‰. In the Gulf of Mexico, there are two sources of brine. The first, halite-derived brines, are derived from the dissolution of salt diapirs. The second, hydrate-derived brines, result from in situ formation of gas hydrates. During hydrate formation, water is removed from the pore fluid to generate gas hydrate, leaving behind residual brine. Hydrate brines differ from halite-derived brine in that they often contain sulfate but they have less ammonium and DOC. Brines are enriched in methane, but also contain high concentrations of ammonium, in the millimolar range, contrasting with lower concentrations of nitrate in the micromolar range (Joye et al. 2010; Bowles et al. 2011, 2016).

In a survey of different brine seeps and briny sediments, higher salinity correlated with increasing ammonium concentrations (Joye et al. 2010). While the relative proportions of NH_4^+ and Cl^- differed locally, the highest ammonium concentrations were consistently associated with the highest local chlorinities, and ammonium concentrations always decreased towards seawater background when chlorinity approached seawater levels (Joye et al. 2010).

Ammonium is thought to originate largely from desorption of particle-bound ammonium that is released into solution when brine comes into contact with subseafloor sediments due the high ionic strength of the brine (Joye et al. 2005; Wankel et al. 2010). Some portion of the ammonium originates from anaerobic decomposition of organic matter that settles into the brines that lie along the seafloor. Brines are also enriched in nutrients such as ferrous iron (Van Capellen et al. 1998), silica, DOC (Joye et al. 2005), and phosphate (Bowles et al. 2016), and thus provide point sources of biolimiting nutrients that enhance microbial growth and activity at the brine interface, as long as increasing salinity and oxidant limitation remain tolerable.

A final factor that controls microbial community and activity is temperature. Cold temperatures imply slow microbial growth and activities in Gulf of Mexico cold seeps. For example, the anaerobic ethane oxidizer "*Candidatus* Argoarchaeum ethanivorans" from Gulf of Mexico cold seeps has been enriched at 12 °C over 10 years (Chen et al. 2019), in contrast to thermophilic, faster-growing alkane-oxidizing archaea from hydrothermally influenced seeps (Laso-Pérez et al. 2016). However, not all cold seeps in the Gulf of Mexico are truly cold in the sense that they lack geothermal input, limiting their thermal regime to the overlying bottom water. Salt formations are thermally more conductive than sediment by a large margin. Salt bodies focus the heat of the natural geothermal gradient in deep sediments, shift isotherms upwards, and warm surrounding sediments by 10–15 °C relative to sediments several km distant, where isotherms can be displaced downwards, indicating large-scale patterns of fluid flow (Bernard-Rayleigh convection cells) driven by the presence of salt bodies (Bennett and Hanor 1987). In consequence, brine seeps and mud volcanoes can have conspicuously elevated temperatures. Thermal pulses up to 48 °C were recorded for an active mud volcano (MacDonald et al. 2000), and a small quiescent brine lake has permanently elevated brine temperature of 17 °C, compared to 7 °C for the overlying seawater (Joye et al. 2005). While these thermal signatures are not comparable to the extremely high temperatures of hydrothermally influenced seeps (Chap. 3), where the thermal limits of life are sometimes exceeded, they would certainly alter the permanently cold seafloor conditions, and permit mesophilic in addition to psychrophilic microbial life.

Microbial Communities at Selected Northern Gulf of Mexico Seep Sites

Seep habitats harbor microbial populations that respond to distinct biogeochemical controls and adapt to specific, sometimes extreme environmental characteristics. A highly varied mosaic of microbial seep habitats and microbial communities often defies attempts at generalization. These environments and their microbial inhabitants are influenced by many interlinking factors, such as the local abundance of methane, alkanes, aromatic compounds, sulfide, the availability of nutrients, salt composition and abundance in different brines, as well as the strength of advective versus diffusional mixing of seep fluids, brines and seawater. For site-specific descriptions, seafloor seeps have to be mapped and identified accurately.

Localizations start with slope regions named for major geological features on the Gulf of Mexico continental slope, extending one degree North-South and one or two degrees East-West (Bryant et al. 1990). These regions are then subdivided into numbered mineral lease blocks of a few square miles each. Seep features are further identified by a combination of regional name and lease block number, for example, GB425 = Garden Banks 425, or GC233 = Green Canyon 233. This chapter will highlight some characteristic seafloor habitats on the northern continental slope of the Gulf of Mexico that are reasonably well documented (Joye et al. 2005, 2009, 2010; Orcutt et al. 2005, 2010; Bowles et al. 2011, 2016; Lloyd et al. 2010; Johansen et al. 2017), and at the same time invite further biogeochemical and microbiological exploration. A few sites of interest are included that have, at least at the time of writing, very little published about them.

This guided tour traverses the northern Gulf of Mexico slope in West-East direction, from Texas to Mississippi (Fig. 1), broadly similar to the track of an extensive survey cruise with R/V *Atlantis* and deep submergence vehicle (DSV) *Alvin* in the fall of 2010 that started in Galveston, Texas, and ended in Gulfport, Mississippi (Expedition AT18-02, Nov. 7 to December 3, 2010; a day-by-day account is available on teskelab2010.wordpress.com). Underutilized seafloor images from this and other research cruises, often taken by *Alvin's* external camera system, are collected here to introduce these fascinating and visually striking seep habitats to a wider audience, and to orientate field researchers who will continue to visit and sample these Gulf of Mexico seeps in the future. Further, biogeochemical and microbiological observations from specific sites that are discussed and summarized here may be widely applicable and characteristic of seep sites on the northern Gulf of Mexico slope (Table 2).

Table 2 Selected seep sites on the northern continental slope of the Gulf of Mexico

Site	Depth (m)	Latitude N	Longitude W	Reference
AC601 brine lake	2340	26°23.46	94°30.90	Roberts et al. (2010b), Crespo-Medina et al. (2016)
AC601 "Red Crater"	2338	26°21.72	94°30.60	Roberts et al. (2010b)
GB425 Mud Volcano	600	27°33.20	92°32.40	Joye et al. (2005)
GC233 Brine Lake NR-1	650	27°43.40	91°16.80	Joye et al. (2005)
GB697 "Hot Site"	1004	27°19.14	92°06.47	Alvin Dive 4546
		27°18.60	92.06.60	Bowles et al. (2016)
GC246 "Dead Crab Lake"	867	27°42.13	90°38.89	Alvin dive 4694
GC600 "Oil Mountain"/ Megaplume 2	1222	27° 22.206	90°34.254	Johansen et al. (2017)
GC600 "Oil Mountain"/ Birthday Candles 2	1222	27° 22.853	90°33.736	Johansen et al. (2017)
MC118 "Woolsey Mound"/ Northwest Crater mat	890	28°51.47	88°29.52	Lloyd et al. (2010)
MC118 "Woolsey Mound"/ Rudyville seep	850	28° 51.136	88°29.531	Johansen et al. (2017)

Positions are given in latitude and longitude with minutes and decimal seconds

Alaminos Canyon 601

The survey starts at Alaminos Canyon 601, a site at the lower slope offshore central Texas that harbors a methane-rich brine lake and an unusually scenic low-activity mud volcano within 2 km of each other. Due to their relatively remote location and greater depth on the outer slope, near the central basin of the Gulf of Mexico, they are not as frequently visited as other more centrally located seeps and brine lakes. The brine lake is located at 2430 m depth; it is approximately circular, has a diameter of ca. 180 m, and an average depth of 4 m (Roberts et al. 2010b). The surface of this brine lake is marked by a readily recognizable, turbid interface with floating barium sulfate (barite) mineral precipitates, which also accumulate on the beach. Barite forms when sulfate-free, barium-enriched brine mixes with sulfate-rich seawater (Roberts et al. 2007). The lake is surrounded by terraced shorelines with small authigenic carbonates that mark higher brine levels in the past (Roberts et al. 2010b).

Sediments close to the shoreline are dark-grey in color, consistent with high porewater sulfide concentrations (mostly between 5 and 10 mM, Crespo-Medina et al. 2016). Sulfide is produced locally by high sulfate-reducing activity in upper halocline sediments at the conspicuous "bathtub ring" of bright barite deposits on the dark-grey beach, and in nearby lower halocline sediments within the shallowest reaches of the lake (Crespo-Medina et al. 2016). A few meters away from the reduced and briny shoreline, surficial sediments become oxidized, and heart urchins and holothurians become abundant, leaving behind a network of trails as they move through the sediment (Fig. 2). This epifauna disappears quickly at a greater distance from the lake. This faunal zonation indicates that these invertebrates derive their livelihood ultimately from the brine lake, by grazing on heterotrophic microbial biomass that thrives on the elevated organic carbon and nitrogen content of beach sediments (in the range of 0.8–1.6 mM DOC, and 0.2–1.8 mM DON), or by grazing on in situ chemosynthetic production that is driven by the brine's elevated content of electron donors, such as ammonia and methane at concentrations >1 mM (Crespo-Medina et al. 2016).

The shore sediments and the edge of the hypersaline shallow brine pool harbor bacterial communities that are shared broadly by cold seeps and general deep-sea sediments, mainly *Deltaproteobacteria, Epsilonproteobacteria* and *Gammaproteobacteria* (Crespo-Medina et al. 2016), whereas obligately halophilic lineages, for example the KB1 bacteria (Nigro et al. 2016) are not found. Similar results were obtained for a brine lake at Green Canyon 233 (surface sample), or in near-surface mud from 25 cm depth in a mud volcano at Garden Banks 425 (Joye et al. 2009). It is possible that the lack of obligate halophiles in these microbial communities reflects the proximity of these sampling sites to the halocline; the microbial communities near the halocline appear to be influenced by marine microbiota growing at standard salinity.

Two kilometers towards the south, we find the mud volcano termed "Red Crater". Named for its extensive red-orange stained mud flows, this area is one of the most impressive seep features in the northern Gulf of Mexico. First documented in 2005 as a large complex of reduced brine flows, reducing dark mud, scattered mussels and fluidized grey mud, the red surface coloration developed subsequently as a likely result of iron oxidation (Roberts et al. 2010b). By the time of a second visit in 2007, the crater then featured an extensive mud flow on its southern rim, and harbored the largest bathymodiolid mussel bed documented on the northern Gulf of Mexico slope on its northwestern rim. Similar mussel beds may have been submerged by episodic mudflows in the past (Roberts et al. 2010b). Red-stained mud flows, extensive mussel beds, and mats of sulfur-oxidizing bacteria were again documented in 2010 (Fig. 2). Red Crater sediment exhibits high sulfate reduction rates and is enriched in methane (near 1.2 mM), sulfide, and ammonium, as is characteristic for mud volcanoes (Table 1, Bowles et al. 2016). In consolidated sediment with siboglinid tubeworms, sulfide and ammonium were recorded in even higher concentrations (7 and 1.8 mM, respectively), compared to fluidized sediments within the mud volcano crater (0.1 and 0.3 mM, respectively). In the latter sediments, seawater is most likely mixing with brine, as indicated by higher sulfate concentrations; this mixing may consequently dilute and oxidize the electron donors derived from the core brine (Bowles et al. 2016). At present, the vertical extent of seawater mixing into the Red Crater, and the geochemical and microbial characteristics of its briny mud source, are unknown.

Garden Banks 425

The Garden Banks region on the upper slope south of the Texas/Louisiana border contains the perhaps best-studied mud volcano in the northern Gulf of Mexico (Fig. 3). In lease block GB425, a large flat-topped mound of approximately 0.7 × 1 km is located at a depth of 570 m on the western edge of the Auger basin (Saager et al. 1999). Near the southwestern edge of the summit plateau, a mud volcano of ca. 50 m diameter emits fluidized, methane-supersaturated, briny mud from a central vent. The fluidized mud overflows the western rim and moves down the southern flank of the mound. The southern rim of the mud volcano appears to be more stable, and is lined by extensive beds of the methanotrophic seep mussel *Bathymodiolus childressi* (MacDonald et al. 2000).

The mud volcano undergoes distinct activity fluctuations; a one-year monitoring time series (July 1997–July 1998) where a temperature probe installed near the central mud vent showed distinct, transient pulses in elevated temperature (max. 48 °C) that coincided with localized oil slicks at the sea surface recorded by satellite-based synthetic aperture radar (MacDonald et al. 2000). Upon recovery, the mud level in the pool, as measured against the thermistor's mooring tether, had risen by ca. 50 cm, indicating gradual accumulation of mud in a potential prelude to major episodic discharges (MacDonald et al. 2000).

This active mud volcano has been profiled biogeochemically and microbiologically, and compared to a small quiescent brine lake (GC233) in the Green Canyon

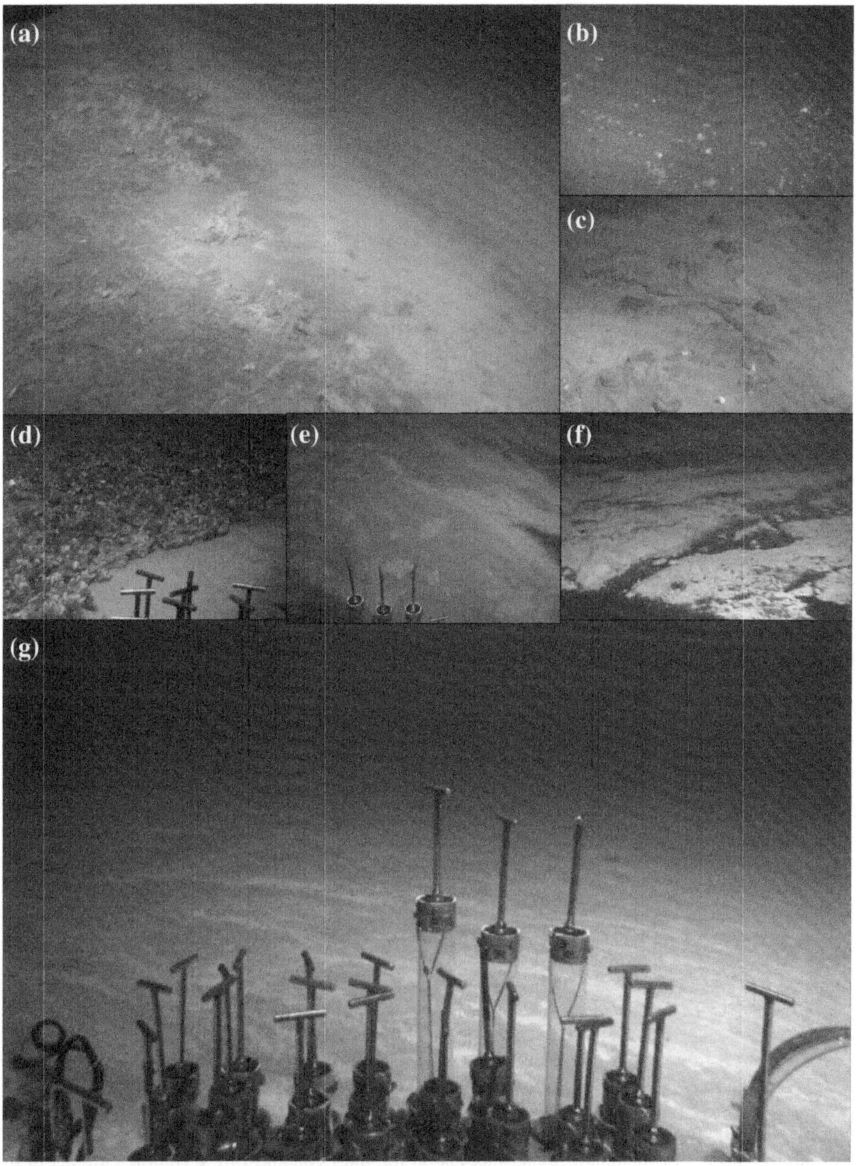

Fig. 2 Alaminos Canyon 601 Brine Lake and mud volcano. **a** From left to right: the brine lake "beach" transitions from dark-grey reducing sediment and barite deposits towards the lighter-colored shoreline and the beginning of the brine basin. **b** The surface of the brine lake with floating precipitates trapped at the seawater/brine interface. **c** Beach sediments with sea urchins and holothurians, and their feeding traces in the sediment. **d** Mussel bed near the large quiescent mud volcano "Red Crater" at Alaminos Canyon. **e** Iron sulfides and oxides color the mud flows at Red Crater; dark streaks indicate reducing conditions, and orange-red coloration indicates oxidizing conditions. **f** Extensive white sulfur-oxidizing microbial mats in the Red Crater area. **g** Panoramic view of Red Crater, with streaking pattern visualizing the direction of slow mud flow. Images from *Alvin* dive 4639, Nov. 10 (**a–c**), and *Alvin* dive 4641, Nov. 12, 2010 (**d–g**)

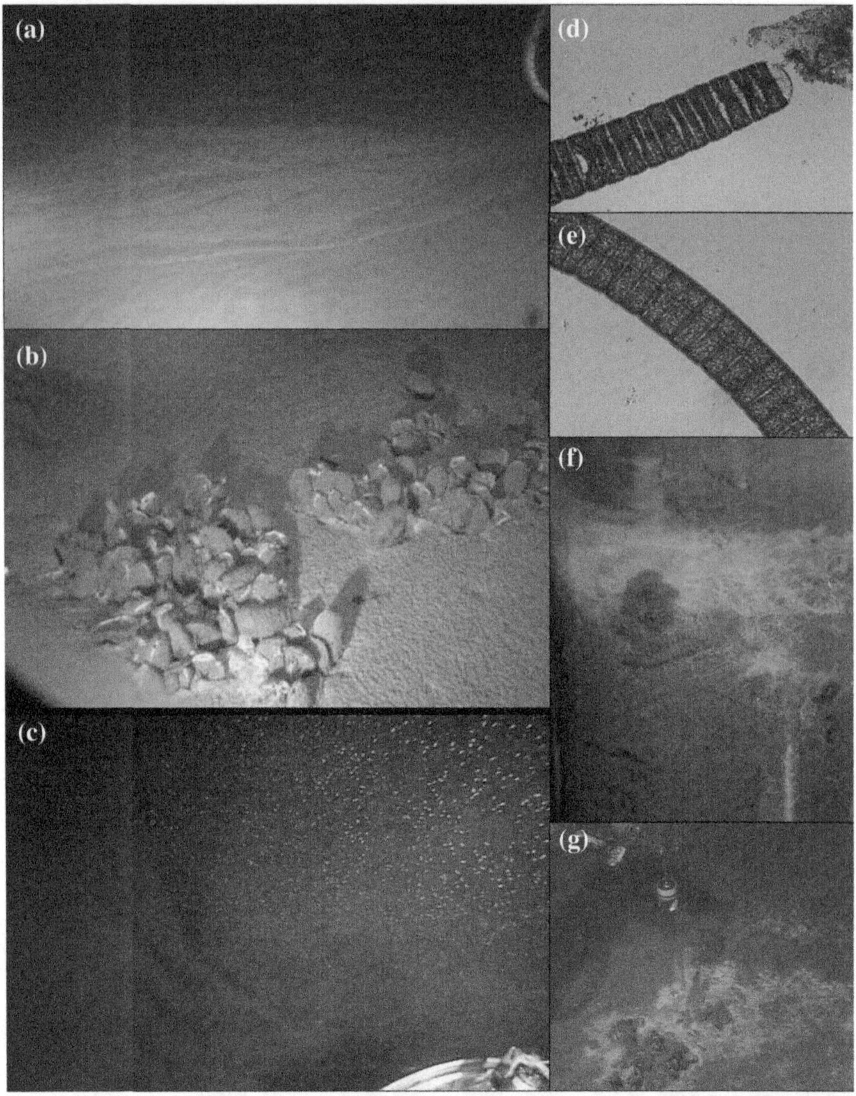

Fig. 3 Garden Banks 425 Mud Volcano. **a** Slowly moving grey mud flows at the GB425 mud volcano. **b** Clusters of methanotrophic mussels on the mud volcano rim. **c** Bubble curtains in the central mud volcano area. **d** and **e** Microphotographs of large, colorless, sulfur-oxidizing filamentous bacteria collected at the GB425 mud volcano; filament diameter is near 90 μm. **f** Microbial mat consisting of large, unpigmented filaments in freshly recovered *Alvin* sediment push core from Alvin Dive 4645. A small pink polychaete worm is visible against the core liner. **g** Core sampling of microbial mat and mussel assemblage on the periphery of the GB425 mud volcano. The push core has an inner diameter of 6.12 cm. Images **a**–**c** from *Alvin* dive 4544, Nov. 14, 2010. Microphotographs **d**, **e** and Photo **f** by A. Teske, UNC Chapel Hill. Image **g** from *Alvin* dive 4645, Nov. 16, 2010

Fig. 4 Green Canyon 233 Brine Lake. The mussel-fringed brine lake at GC233. **a** In situ camera placed on the massive mussel bed surrounding the GC233 brine lake, visible as dark blue void in the background. *Alvin* Dive 4654, Nov. 25, 2010. **b** Sediment push coring of orange filamentous bacteria near the outer edge of the mussel bed during *Alvin* dive 4653, Nov. 24, 2010. **c** Sediment core from this sampling trip with orange filamentous bacteria. **d** Individual filament, average diameter 38 μm. Images **a** and **b** courtesy of Alvin team and WHOI; images **c** and **d** by A. Teske, UNC Chapel Hill

area further east (Fig. 4) to assess geochemical and microbiological differences between the two habitat types (Joye et al. 2009). Due to the discharge of fluidized mud and sulfate-depleted brine, the salinity and sulfate gradients at the seawater interface of the mud volcano were steeper than at the brine lake. The transition zone where seawater and brine mix extended ca. 60 cm into the mud volcano, and ca. 120 cm into the brine lake. Sulfate reduction rates in the mud volcano remained approximately one order of magnitude lower than in the brine lake (60 μM d^{-1} vs. 5 μM d^{-1}), and sulfate-reducing activity was limited to the mixing zone at the mud volcano surface (Joye et al. 2009).

Based on PCR amplification and clone library sequencing of 16S rRNA and *dsrAB* genes, the microbial community of the mud volcano included fewer clades of sulfate-reducing *Deltaproteobacteria* and sulfur-oxidizing *Epsilonproteobacteria* than the brine lake (Joye et al. 2009). Due to the limitations of clone library coverage, conclusions about the different relative importance of sulfate-reducing and sulfur-oxidizing bacteria in the two habitats have to be treated with caution; however, the sequence data identify shared community members. For example, the mud volcano and the brine lake both yielded clones of the *Desulfobacterium anilini* clade, now renamed *Desulfatiglans anilini* (Suzuki et al. 2014), whose cultured representatives specialize in the complete oxidation of aromatic hydrocarbons (Teske 2019). Obviously, the hydrocarbons in the brine and in the mud volcano select for specialized hydrocarbon-oxidizing sulfate-reducing communities.

Given the elevated salinity in both habitats, a thorough isolation and characterization effort should yield new halophilic or halotolerant species and genera of anaerobic hydrocarbon oxidizers. After sulfate was consumed at approximately 50 cm depth in the mud volcano, methanogenic activity increased conspicuously (Joye et al. 2009). Acetoclastic methanogenesis rates in the mud volcano exceeded these rates in the brine lake by two orders of magnitude (max. 120 vs. 0.3 nM d^{-1}); hydrogenotrophic activity was not detectable in the mud volcano. However, methanogenesis rates were not determined for non-competitive substrates that are not commonly used by sulfate reducers, such as methylamines and methanol, leaving open the possibility of higher integrated methane production rates in these brines. The sequencing results indicate potential for methylotrophic methanogenesis. In the brine lake and in the mud volcano, *mcrA* genes for uncultured members of the *Methanosarcinales* were found, which often specialize in using methylated compounds as methanogenic substrates (Liu and Whitman 2008). Recent work has shown that methylotrophic methanogenesis is often the preferred methanogenic process in marine sediments (Zhuang et al. 2016).

Garden Banks 697

The GB697 lease block, approximately 30 miles southeast of the GB425 mud volcano, is characterized by gas and oil seepage, and active salt tectonics. Here, a small mud volcano, also known as the "Hot Site", vigorously discharges gas and fluidized briny mud (Fig. 5). This site is far more active than the GB425 site. Sediments around the discharging mud volcano exhibited a pore fluid salinity of ~160. The temperature of the discharging fluid was higher than the ambient bottom water (~20 °C). Discharging fluids rapidly mixed with seawater, diluting the salinity, but still, the fluids exhibited high concentrations of methane (>500 μM). Patchy microbial mats grew atop the briny mud flow. Although this mud volcano was visited by R/V *Atlantis* and DSV *Alvin* in 2010 and 2014, published site descriptions are still missing. Porewater data from briny sediment cores collected in 2007 show the classical brine signature of elevated methane (>300 μM), ammonium (>2000 μM), phosphate (10 μM), and DOC (>1500 μM) concentrations (Joye et al. unpublished data). Cores collected from around the venting mud volcano showed consistent brine impact, suggesting consistent brine discharge in the area.

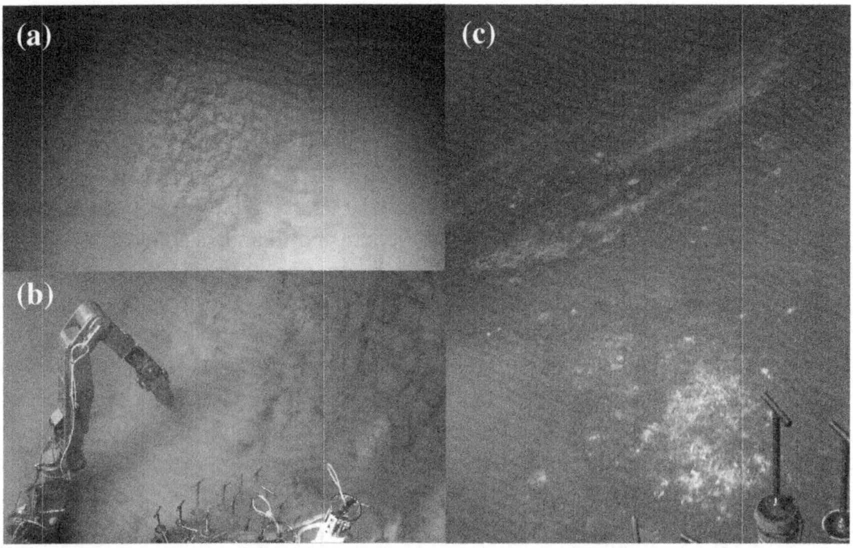

Fig. 5 Garden Banks 697 Mud Volcano. **a** Briny mudflow at the mud volcano; due to the density difference, vigorous convection does not obscure the interface between seawater and the fluidized briny mud. **b** Sediment sampling in the briny mudflow flowing downhill from the mud volcano crater; the base of the sampling arm and the partially obscured core quivers are ca. 1 m above the solid bottom. **c** Abundant microbial mats outside of mudflow area. Images **a** and **b** from *Alvin* dive 4701, April 20, 2014; Image **c** from *Alvin* dive 4546, Nov. 17, 2010

Green Canyon 246

We complete our survey of understudied mud volcanoes in the Gulf of Mexico by introducing Dead Crab Lake, located in Green Canyon 246. Some brine lakes represent quiescent mud volcanoes that have not erupted for hundreds of years. For example, the mussel-fringed brine lake at GC233 sits like a small caldera lake on top of an apparently quiescent mud volcano, where stratified sediments layers indicate distinct episodes of activity over the last 15,000 years (MacDonald and Peccini 2009). Dead Crab Lake demonstrates that a small brine lake can resume activity and revert to a mud volcano. During a visit with R/V *Atlantis* and DSV *Alvin* in November 2010, Dead Crab Lake (ca. 10 m diameter) was filled with dark halite-saturated brine. The lake was surrounded by extensive brine flow areas and reduced sediments with orange, grey and black surface coloration due to a unique cocktail of metals in variable oxidation states. The southern shore of the lake featured small ($\sim$15 cm tall) barite chimneys distributed among orange-streaked mud flows—similar to previous observations on a much larger scale at the "Red Crater" in Alaminos Canyon (Fig. 6). Reducing sediment patches to the west and northwest of the lake edge were overgrown with tuft-like microbial mats of non-pigmented, large sulfur-oxidizing filaments, later identified as members of the candidate lineage "Marithrix" within the family *Beggiatoaceae* (Salman-Carvalho et al. 2016, and Chap. 8). The sediment cores from the "*Ca.* Marithrix" mat location were saline (up to 100 PSU), rich in DOC and DIC (around 0.6 mM, and 5–7 mM, respectively), strongly enriched in ammonium (up to 1.8 mM), and highly sulfidic with a peak near 4 mM at 4 cm depth, but nearly sulfide-free in the surficial centimeter layer (Salman-Carvalho et al. 2016).

Filamentous sulfide-oxidizing bacteria at Dead Crab Lake were also studied with regard to their fossilization potential. Barite-encrusted filaments showed 16S rRNA gene signatures related to "*Candidatus* Maribeggiatoa" (Stevens et al. 2015). By oxidizing sulfide to sulfate, these organisms contribute to their own encrustation by barite, when inundated by barium-rich, sulfate-depleted brine fluids. This microbially-catalyzed barite encrustation process would likely compete with abiotic barite precipitation upon seawater exposure (Stevens et al. 2015).

While all microbial and geochemical observations during the 2010 visit indicated a reducing brine lake surrounded by brine flows, colorful precipitates of metal oxyhydroxides, and patches of filamentous large sulfur-oxidizing bacteria, these conditions changed over the next four years. During the second visit with DSV *Alvin* in April 2014, the lake basin was almost entirely filled with light-grey mud, with small circles along the smooth mud surface indicated rising and discharging bubbles of methane-saturated fluidized mud. The slope on the northeastern shore and the liquefied mud in the lake shared the same light-grey color, which contrasted with the darker olive-grey and brown color of the surrounding seafloor sediment. The northeastern shoreline of the mud lake, where the liquefied mud met the slope surrounding the lake basin, was marked by a narrow zone of dark-grey residual reducing brine (Fig. 6e). To the northwest and west of the lake, the lake is not contained by a slope, but gradually spreads in a network of shallow channels and rivulets containing reducing brine over surficially oxidized sediments. Beds and clusters of small bathymodiolid mussels had settled in this area since 2010 and were proliferating in 2014.

The interplay of brine and sediments creates an extensive and intricate mosaic of orange-rust colored and dark-grey seafloor (Fig. 6f). The colors were not quite as flamboyant as the intense orange seafloor hues that were observed sometimes in 2010, and initially mistaken for microbial mats (Fig. 6a). Occasionally, eruptions are sending rivulets and clouds of finely suspended sediment over the lake's slope and onto the surrounding seafloor, where they deposit a fresh coat of light-grey sediments covering the darker, partially reducing seafloor (Fig. 6g and h). Sediment cores collected with DSV *Alvin* at some distance from the lake during an enforced "mud cloud break" in 2014 contained sheathed bundles of filaments within the upper 2–3 cm of oxidized, non-sulfidic sediment; these sheathed filament bundles resembled the sulfur-oxidizing, sediment-dwelling filamentous bacterium *Thioploca* (Chap. 8). The small barite chimneys on the southern shore of the lake were active in 2014, just as they were in 2010, emitting tiny mud clouds to great decorative effect (Fig. 6i).

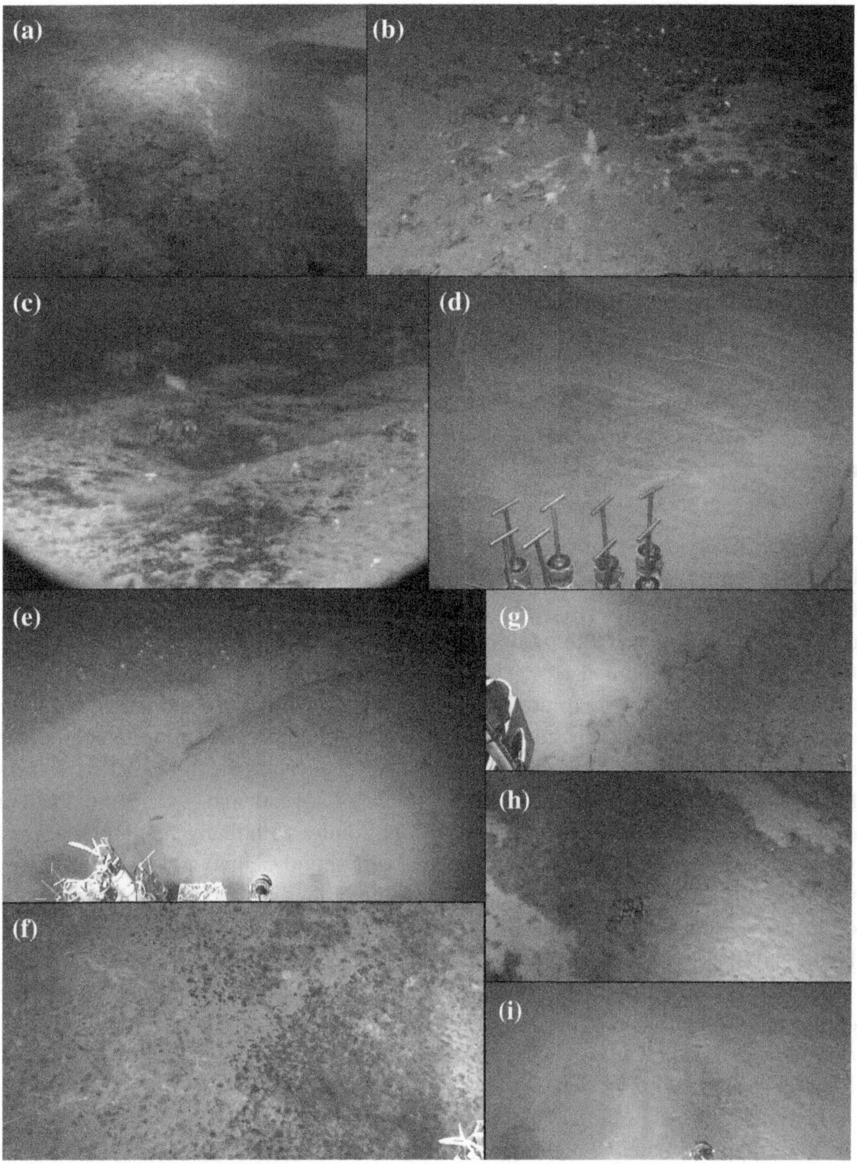
(a)
(b)
(c)
(d)
(e)
(g)
(h)
(f)
(i)

◂**Fig. 6** **Dead Crab Lake at GC246.** **a** Orange-stained brine flow area where seawater exposure oxidizes iron in the brine fluids. **b** Small barite (barium sulfate) chimneys in an area with reducing sediments, indicating precipitation from sulfate-free brine fluids at exposure to seawater. **c** Small low-lying seafloor area with reducing brine and chemosynthetic mussels. **d** Miniature mud volcanoes and mud flows, streaked with lightly orange and rust-toned iron oxides, from the southern end of Dead Crab Lake. **e** The mud-filled lake basin and its northeastern slope in 2014. **f** Characteristic mosaic of oxidized, rust-colored and reduced, dark-grey and black sediments west and southwest of the lake; small grey brine flow channels are cutting across in the oxidized sediments. **g** A cloud of finely dispersed particles moves from the direction of the lake towards *Alvin*; it cleared within minutes, but reveals its origin in a fresh light-grey dusting of the lake slope. **h** Fresh flow of light-grey sediment contrasts with partially reducing seafloor, with mussel cluster at center. **i** A small mud volcano at the southern end of the lake emitting a miniature eruption, a puff of suspended light particles ca. 10 cm high. Images **a** and **b** from *Alvin* dive 4562, Nov. 23, 2010; Images **c** and **d** from *Alvin* dive 4561, Nov. 22, 2010; Images **e–i** from *Alvin* dive 4694, April 10, 2014

Green Canyon 600

The Green Canyon 600 area on the upper continental slope of the Gulf of Mexico harbors a seafloor ridge with the most productive natural oil seeps in the northern Gulf of Mexico, also known as Oil Mountain. Here, oil and gas bubbles rising from the seafloor at 1200 m depth produce prolific oil slicks at the sea surface that extend for tens of kilometers, and are visible from space (Garcia-Pineda et al. 2010). The GC600 area was initially explored by DSV *Alvin* dives in 2006, and turned out to be structurally and biologically complex. A NW-SE trending salt-supported carbonate ridge between two intraslope basins is highly faulted and fractured on its eastern flank. These faults provide a migration pathway for oil and gas to the seafloor (Roberts et al. 2010b). The rough seafloor topography with many small mounds and valleys harbors patches of microbial mats, vesicomyid clams, and vestimentiferan tubeworms. Gas bubble streams emerge through cracks in carbonate pavement, and oil droplets seep from the sediment in low-lying pockmarks (Roberts et al. 2010b). Hydrocarbons stain surface-breaching gas hydrates dark brown, so that partially exposed seafloor hydrates stand out against the lighter-colored sediment cover. The exposed surface area of gas hydrates harbors pink ice worms (*Hesiocaeca methanicola*) nestled into small depressions along the hydrate surface (Fig. 18 in Roberts et al. 1990; Fig. 4 in Johansen et al. 2017; Fig. 7).

Hydrocarbons are emitted in conspicuous bubble plumes (Fig. 7) at two major locations approximately 1 km distant from each other, the Mega Plume site on the northwestern ridge and the Birthday Candles site on the southeastern ridge (Johansen et al. 2017). The Birthday Candle bubbles consist of liquid hydrocarbons; they are emitted at lower rates compared to the Mega Plume bubbles that consist of a gas/liquid hydrocarbon mixture. Interestingly, both bubble types are larger than the gas bubbles that emerge at high venting rates at another location, the Rudyville seeps at Mississippi Canyon 118 (average diameters of 5 and 3.9 vs. 3 mm, respectively). Thus, the greater viscosity of liquid hydrocarbons compared to methane gas favors larger bubble size, while slowing down the discharge rate

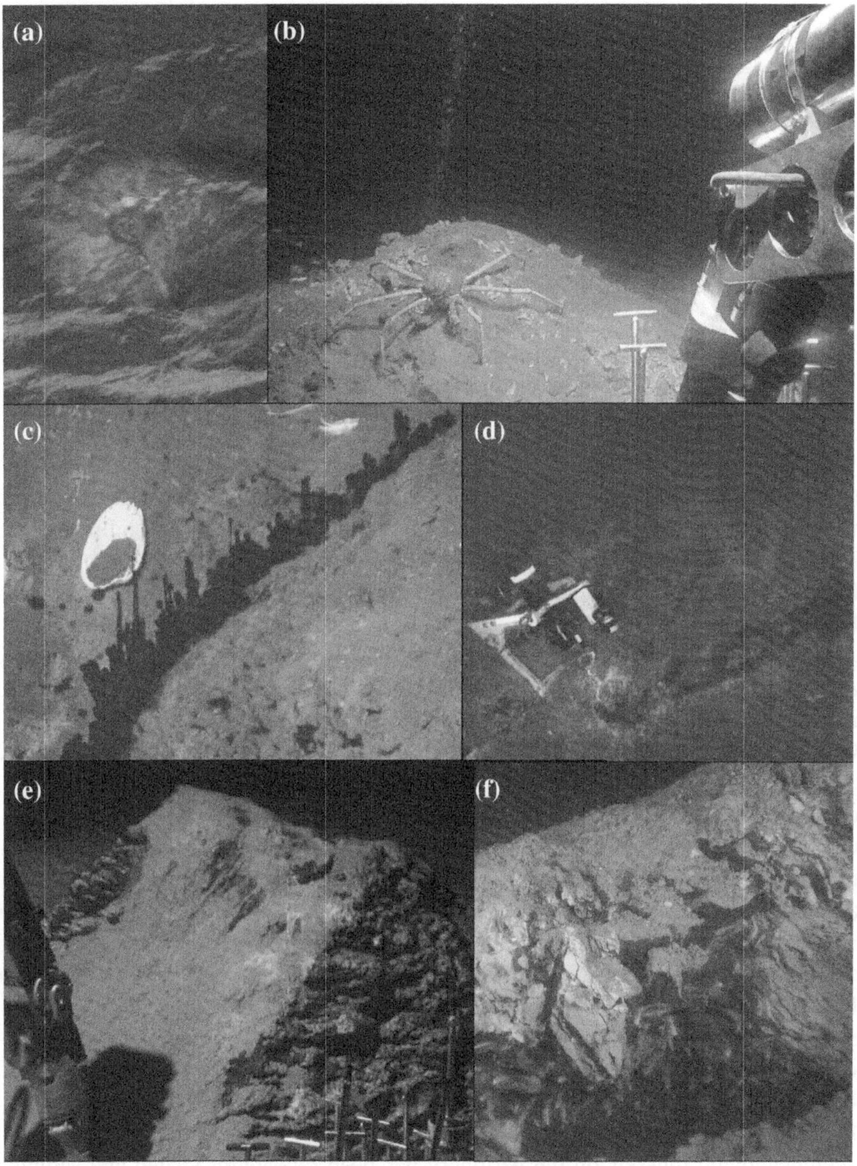
(a)
(b)
(c)
(d)
(e)
(f)

◂**Fig. 7** **Oil Mountain at GC600.** The image panels illustrate and re-trace hydrocarbon flow from seafloor seepage to sea surface at Oil Mountain (see also Teske 2019). **a** An oil droplet is spreading in a rainbow-colored oil sheen on the sea surface after rising from the seafloor at ca. 1200 m depth; these oil sheens are prominently visible on calm days. **b** A sediment-covered hydrate mound with shell hash and resident crab releases a gas bubble stream at its top. Photo *Alvin* Dive 4655, Nov. 26, 2010. **c** Short strings of dark-brown viscous hydrocarbons emerge from the sediment near the base of a petroleum-stained hydrate mound at the "Birthday Candles" site. Oil droplets detach periodically from the candle tips and rise into the water column. Photo *Alvin* dive 4689, April 3, 2014. **d** An automated seafloor camera is recording a stream of oil droplets emerging from sediment at the Mega Plume site. *Alvin* dive 4690, April 4, 2014. **e** A massive hydrate outcrop, ca. 2 m high, is viewed against *Alvin's* sampling arm. *Alvin* dive 4690, April 4, 2014. **f** The same outcrop harbors ice worms occupying crevices on the brown-stained hydrate. Cracks in the light-grey sediment cover reveal the dynamics of the underlying hydrate. Image **a**: A. Teske, UNC Chapel Hill; Images **d** and **f**: I. MacDonald, Florida State University; other photos taken by *Alvin* camera system. The photo in **e** was color-manipulated to reduce the exaggerated blue color tone of the original

(Johansen et al. 2017). It appears reasonable that these interconnected differences in venting rate, bubble size, and hydrocarbon composition should influence the composition and activity of hydrocarbon-degrading microbial communities. Microbial mats and invertebrate fauna are consistently associated with these seep locations (Johansen et al. 2017).

The bubble streams do not only impact benthic processes along the seabed, they also entrain deep water towards the surface, causing localized upwelling and increasing water column productivity above the seep sites. Multi-year surveys at GC600 showed that the surficial water column shows twofold higher chlorophyll concentrations, centered on a depth of 82 m, compared to background chlorophyll concentrations that peaked at 99 m depth (D'souza et al. 2016). Further, significantly elevated nitrate/nitrite concentrations were shifted upward in the water column relative to the background profiles (D'souza et al. 2016). Elevated chlorophyll and nitrate/nitrite concentrations in the upper water column co-occurred with cooler temperatures, which, in combination, indicates upwelling of cool, nutrient-enriched deep water at seep sites that stimulates phytoplankton growth in more shallow water layers (D'souza et al. 2016).

Oil Mountain harbors additional seafloor features that have not yet been documented in the literature. Hydrates exclude salts during their formation, and form localized brine flows at the seafloor that do not derive from dissolution of deep salt formations or salt domes. A small brine flow with patchy microbial mats and chemosynthetic mussels on the periphery has been repeatedly visited and sampled during Expedition AT26-13 with R/V *Atlantis* and DSV *Alvin* in 2014. The meandering, narrow flow channel that ends in a small basin inspired the name "Cobra Brine Lake" for this location (Fig. 8). These hydrate-derived brines differ from halite-derived brines. Namely, hydrate-derived brines contain more sulfate and are more sulfidic, and their salt composition reflects their seawater origin relative to halite-derived brines that result from clay dewatering and halite dissolution (Haeckel et al. 2004; Torres et al. 2004).

Fig. 8 Brine flow at GC600. a The lower basin of an extended brine flow channel, also referred to as "Cobra Brine Lake", is filled with dark, sulfidic brine flow and contrasts strongly against the olive-grey seafloor sediments. **b** A close-up view of the seep community includes shell hash, diverse crabs, and sulfur-oxidizing mat near the edge of the brine flow. The two red laser dots mark a distance of 10 cm. **c** Dense clusters of seep mussels are intercepting reducing fluids at the edge of the brine channel. **d** The "Cobra Brine" reducing sediments form a grey-whitish halo of sulfur precipitates that is surrounding the dark brine. All photos taken by *Alvin* camera system during dive 4696

Mississippi Canyon 118

Located on Louisiana's continental slope at a depth of ca. 880–900 m, Mississippi Canyon lease block 118 (MC118) harbors an extensive complex of carbonates and hydrates with numerous cold seeps, vents, and seafloor pockmarks, termed Woolsey Mound (Macelloni et al. 2013). Woolsey Mound became one of the most extensively studied sites on the Gulf of Mexico continental slope after the implementation of a seafloor observatory to monitor the massive gas hydrate mounds along the seabed (McGee 2006). Thus, the Woolsey Mound hydrates, seeps and microbial mats are among the best-studied features on the northern Gulf of Mexico slope (Fig. 9).

The exposed hydrates at MC118 and at a second site offshore British Columbia have been studied extensively to understand the chemical, physical and biological controls that stabilize these hydrate outcrops over many years of observation, with

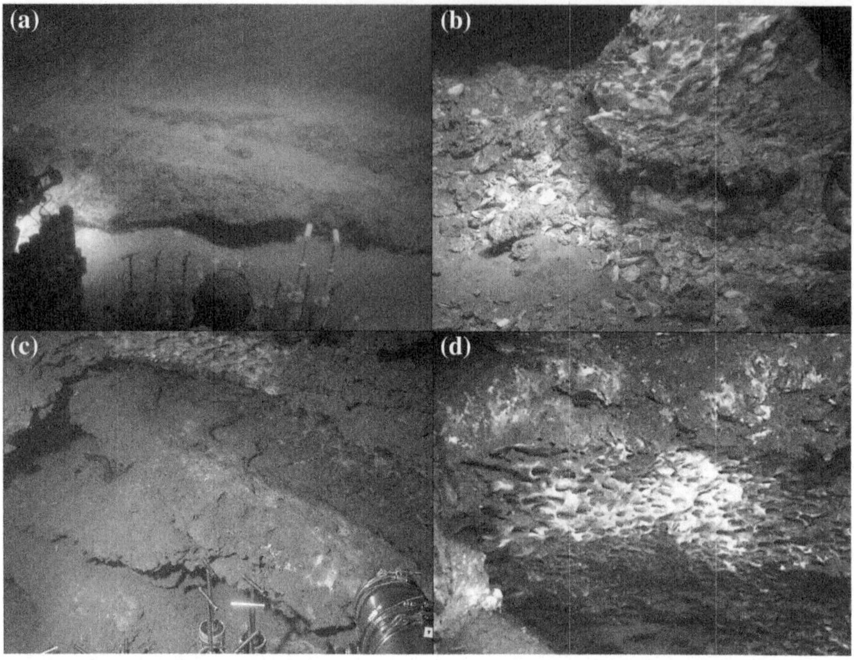

Fig. 9 Hydrates at Woolsey Mound, MC118. a The largest seafloor hydrate outcrop in the Gulf of Mexico, "Sleeping Dragon", in the southwestern complex of MC118. The hydrate outcrops are covered with a sediment blanket. **b** Exposed petroleum-stained seafloor hydrate at the Rudyville location in the southwestern complex of Woolsey Mound, surrounded by abundant shell hash and broken rough-textured carbonate concretions. **c** Multi-tiered, partially sedimented "galleries" of exposed seafloor hydrates near the base of a hydrate outcrop where the bulk of the hydrate emerges from the surrounding seafloor sediments; the two red laser dots mark a distance of 10 cm. Bottom-dwelling fish at center left. **d** View of "Sleeping Dragon" hydrate, showing the contrast between yellow, petroleum-stained hydrate with ice worm cavities at the base of the emerging outcrop, and the dark-grey sedimented carbonate cap. Image **a**: Joye Research Group, Univ. of Georgia; Images **b** and **c**: *Alvin* dive 4657, Nov. 28, 2010; Image **d**: Mississippi Mineral Resources Institute, University of Mississippi

only minimal erosion of the exposed hydrate surface (Lapham et al. 2010). In situ concentration measurements of methane in the sediments surrounding the hydrates have shown that porewater methane concentrations are much lower than the equilibrium concentrations that would be required to maintain equilibrium with gas hydrates (Lapham et al. 2010). Methane concentrations in the thin sediment blanket on top of the hydrates never exceeded 3 mM, compared to a predicted value of ca. 70 mM methane at equilibrium concentration (Lapham et al. 2010). However, pore fluid methane concentrations near the Bullseye Vent (Cascadia Margin) exceeded 80 mM, showing that saturated and even supersaturated methane concentrations can exist near areas of active hydrate formation (Lapham et al. 2013). Anaerobic methane oxidation changed the $\delta^{13}C$ signature of porewater methane from −50 to

−35‰ with greater distance from the hydrate, indicating that an increasing proportion of the porewater methane had been oxidized. Yet, quantitative models showed that this methane-consuming microbial process in itself would not have been sufficient to explain the difference between predicted and measured porewater methane concentrations. A number of proposed mechanisms could protect and stabilize gas hydrates, including surface stabilization by oil coatings (the MC118 gas hydrates are yellow- and beige-colored due to oil content), or, by microbial biofilms of methane- or oil-degrading bacteria that grow on the hydrate surface (Lapham et al. 2010).

Acoustic surveys of the seafloor at Woolsey Mound have identified numerous wipeout zones, and sediment cores from these areas showed extensive methane seepage and sulfate reducing activity (Lapham et al. 2008). Three main crater complexes are associated with distinct characteristics: while the southeastern complex appears to be extinct or quiescent, the southwestern complex is fully active and shows oil and gas venting, and the northwestern complex represents a young stage of gas seepage and venting (Macelloni et al. 2013). Microbial activity in sediment cores from these three complexes support this picture of a mound with evolving and changing seepage patterns: cores from the northwestern and southwestern area show high methane-oxidizing and sulfate-reducing activity, whereas cores from the southeastern complex show low activity (Lapham et al. 2008). Sulfate reduction rates at oil seeps in the southwestern complex were driven predominantly by oxidation of petroleum hydrocarbons, including the abundant aliphatic hydrocarbons that were found in concentrations of mg/g wet sediment. Methane was an insignificant substrate for microorganism at the oily seeps, and is assumed to be of greater importance at the gassy seeps in the northwestern complex, where aliphatic hydrocarbon abundance was reduced by one or two orders of magnitude (Bowles et al. 2011). At both oily and gassy sites, hydrogen concentrations were likely controlled by sulfate reduction (Lin et al. 2012), and remained at low nanomolar concentrations, mostly between 5 and 20 nM, within a total range of 1.6–57.9 nM (Bowles et al. 2011). Despite differences in relative substrate importance at the oily and gassy sites, clone library analyses showed similar populations of methane-oxidizing archaea (ANME-1, ANME-2 subgroups) and sulfate-reducing bacteria (*Desulfosarcinales*, *Desulfatiglans* cluster, and uncultured lineages) throughout these sampling sites (Bowles et al. 2011). Specific clusters within the *Desulfosarcinaceae* and *Desulfobulbaceae* participate syntrophically in methane oxidation (Knittel et al. 2003; Schreiber et al. 2010), whereas diverse cultured *Desulfosarcinales* and all currently known members of the *Desulfatiglans* cluster use aliphatic and aromatic hydrocarbons as substrates (Teske 2019).

The greater diversity of hydrocarbon-oxidizing sulfate reducers compared to methane-oxidizing syntrophic specialists suggests that sulfate reduction rates at hydrocarbon seeps are uncoupled or only weakly coupled to methane oxidation since other hydrocarbons and organic substrates are commonly available (Bowles et al. 2011, 2019). Interestingly, at a natural seep site, all other substrates would have to be consumed before sulfate reduction would be fueled by methane oxidation. An extensive meta-analysis of published field studies showed indeed that methane oxidation sustains only a fraction of measured sulfate reduction rates, with a median ratio of sulfate reduction to methane oxidation rates of 10.7, based on highly variable datasets from 52 different seep sites within and beyond the Gulf of Mexico (Bowles et al. 2011). The uncoupling of sulfate reduction and methane oxidation appears to be particularly evident at the haloclines of seafloor brine lakes, where abundant organic substrates fuel high rates of sulfate reduction and diminish the importance of methane as sulfate-reducing substrate (Crespo-Medina et al. 2016).

Woolsey Mound features extensive gas seeps and associated sulfur-oxidizing microbial mats in the emerging northwestern seep complex. These mats occur in extensive pockmarks, which may represent the result of recent venting episodes. The absence of chemosynthetic invertebrates in this area suggests that stable seepage conditions have not yet been established, or, have not prevailed long enough to allow colonization by seep epifauna (Macelloni et al. 2013). Mat-covered seep sediments at MC118 have been studied with respect to how microbial community composition, diversity, and biogeochemical activity change from the center of the mat to its margins (Lloyd et al. 2010). Within the small spatial scale of a seep hot spot (a few meters or less), the microbial community structure and activity transitions between seep-adapted sulfate-reducing bacteria and methane-oxidizing archaea that rely on fossil hydrocarbons towards a more diverse microbial community where these processes and populations are increasingly embedded, and are finally supplanted by complex marine benthic microbiota that ultimately depend on degrading and assimilating photosynthetic biomass that originates in the photic zone (Lloyd et al. 2010).

A mat-covered area in the northwestern complex of Woolsey Mound was investigated and sampled during dives with DSV *Johnson Sealink* in April 2006. Here, the upper 5 cm of the sediment showed high rates of sulfate reduction (max. 250–500 μM sulfate reduced per day) and methane oxidation (max. 50–100 μM methane oxidized per day); the rates declined quickly in deeper sediment layers (Lloyd et al. 2010). Also, below the upper 5 cm porewater sulfate concentrations decreased to <1 mM, while porewater methane concentrations (dominated by biogenic methane, $\delta^{13}C$ in the range of 45–55‰) were regularly near or above 1.5 mM. Consistent with the geochemical data, the analysis of reserve-transcribed 16S rRNA in these sediments showed that the bacterial populations were dominated by different deltaproteobacterial sulfate reducers (members of the SEEP-SRB1 group, diverse *Desulfobacteraceae*, and the uncultured SEEP-SRB2 group, a seep specialist within the *Desulfobulbaceae*) and by anaerobic methane oxidizing archaea of the ANME-1, ANME-2a and ANME-2c groups (Lloyd et al. 2010).

Transcriptional analyses of the *dsrAB* and *mcrA* genes yielded predominantly members of an uncultured *dsrAB* cluster, and of ANME-2 lineages (Lloyd et al. 2010). *mcrA* genes of ANME-1 archaea were presumably underreported since the commonly used *mcrA* primer combination at that time selected against ANME-1 (Lever and Teske 2015). These active bacterial and archaeal populations changed towards the bare sediments outside of the sulfate-reducing and methane-oxidizing mat-covered sediment, containing a diverse assemblage of *Gamma-* and *Alphaproteobacteria*, *Bacteroidetes*, *Chloroflexi*, *Gemmatimonadetes*, *Planctomyces*, *Acidobateria*, and *Thaumarchaeota*. Consistent with the decline in sulfate reduction and methane oxidation rates, *dsrAB* and *mcrA* gene transcripts could no longer be detected (Lloyd et al. 2010). Interestingly, this change in microbial community structure and activity towards the mat margin was evident in surficial sediment (0–3 cm), but appeared attenuated in deeper sediments where the 16S rRNA and functional gene transcription patterns of the seepage-dominated, mat-covered sediment continued to a large extent at the margin of the mat (Lloyd et al. 2010). In other words, the microbial seep community persisted a few centimeters below the sediment surface. Pyrosequencing of the V6 region of 16S rRNA transcripts identified the highest bacterial diversity in surficial as well as shallow sub-surface sediments outside of the mat, consistent with the notion that seep conditions reduce bacterial diversity and select for a limited set of specialists. Archaeal diversity peaked in the shallow subsurface (but not the surface) outside of the mat area, as the predominance of the *Thaumarchaeota* may have reduced archaeal diversity in the surficial sediment (Lloyd et al. 2010).

At MC118, mat-associated microbial communities are dominated by deltaproteobacterial sulfate reducers, including hydrocarbon-degrading taxa, and by methane-oxidizing ANME archaea. These findings match the results from other cold seep microbial mats in the Gulf of Mexico (GC185, also referred to as "Bush Hill", and GC234) where cloning and sequencing of reverse-transcribed rRNA from orange and white *Beggiatoaceae* mats and their underlying sediments yielded predominantly *Deltaproteobacteria*, in particular members of the *Desulfosarcinales*, the *Desulfatiglans* group, the *Desulfobulbaceae*, and diverse ANME-2 archaea (Mills et al. 2004). Phylotypes of the visually dominant mat-forming *Beggiatoaceae* were found only in low abundance.

Some bacterial and archaeal populations that dominated in transcript analyses of microbial mat-associated shallow sediments extend deeper into the sedimentary subsurface. In a DNA-based sequencing survey of a 4.5 m long gravity core from the northwestern edge of the northwestern complex at MC118, deltaproteobacterial seep specialists (SEEP SRB-1, SRB-2) were found in the sulfate-reducing zone, and even at 2 m depth (SEEP-SRB-2) below the methane/sulfate interface at 1.2 m (Underwood et al. 2016). DNA-based 16S rRNA and *mcrA* gene sequencing detected ANME-1 archaea as the dominant archaeal population throughout the entire core. Taken together, the occurrence of the ANME-1 archaea in methanogenic sediments below the methane-sulfate interface, the apparent lack of methanogenic archaea in these sediments, and the absence of sulfate-reducing ANME syntrophs (Underwood et al. 2016) supports the hypothesis that ANME-1 archaea can also survive as methanogens (Lloyd et al. 2011; Bowles et al. 2011; Orcutt et al. 2005).

While microbial methane and sulfur cycling at hydrocarbon seeps has received much attention, nitrogen transformations—essential for any microbial ecosystem—have just begun to be studied. Conspicuous, large sulfur bacteria of the family *Beggiatoaceae* form extensive, easily visible microbial mats at seep sites (Teske and Salman 2014), and many types of marine *Beggiatoaceae* accumulate the electron acceptor nitrate intracellularly, reducing it to dinitrogen or ammonia (McHatton et al. 1996; Otte et al. 1999). Therefore, mats of sulfur-oxidizing bacteria and the underlying sediments were selected for measurements of denitrification rates at MC118 (Bowles and Joye 2011). Measurements of high nitrate reduction rates in surficial mat sediments (0–6 cm, 32 μM d^{-1}), and lower rates in the filament-free sediment horizon underneath (6–12 cm, 11 μM d^{-1}) translated into depth-integrated rates of 80 and 27 μM nitrate reduced m^{-2} h^{-1} for the two sediment horizons, similar to rates from coastal marine sediments and moderately eutrophic freshwater environments (Bowles and Joye 2011). These high nitrate reduction rates should not be ascribed to the mat-forming *Beggiatoaceae* alone; it is very likely that other members of the highly complex mat-associated bacterial and archaeal communities (Mills et al. 2004) are contributing significantly to the process. Nitrate-reducing hydrocarbon-oxidizing bacteria (Widdel and Rabus 2001) are particularly interesting candidates for linking nitrogen cycling with fossil carbon assimilation and remineralization in Gulf of Mexico cold seeps.

Mississippi Canyon 252

Located on the Gulf of Mexico continental slope south of Mississippi, MC252 is the site of the Deepwater Horizon drilling platform explosion (20 April 2010) that led to the release of 4.1×10^6 barrels of liquid oil (McNutt et al. 2012), and ca. 1.7×10^8 kg of gas (Reddy et al. 2012) over the next three months until the wellhead could be sealed in July 2010. This massive discharge of hydrocarbons into the Gulf of Mexico fundamentally differs from the gradual seepage characteristic of natural hydrocarbon seeps, not only by quantity alone, but also by the nature of hydrocarbon delivery that reached the seafloor in a massive pulse of hydrocarbon fallout and sedimentation from the upper water column. To highlight these differences, we conclude this overview on Gulf of Mexico seeps with a short summary of the Deepwater Horizon impact on the seafloor and its microbial communities, and seafloor observations near the Macondo wellhead during the *Alvin* survey in November 2010. In conspicuous contrast to the abundant benthic invertebrates and microbial mats at hydrocarbon seeps throughout the Gulf of Mexico, the Deepwater Horizon oil sedimentation pulse had devastating impacts on deep-sea fauna (Joye et al. 2019) and reduced the seafloor near the wellhead to a wasteland of dead worms and pteropod shells (Fig. 10).

Within a few days of the wellhead explosion, large amounts of marine oil-derived snow formed at the sea surface (Passow et al. 2012; Ziervogel et al. 2012). These floating aggregates were composed of hydrocarbon droplets trapped within a sticky matrix of microbially produced extracellular polymeric substances, which became the nucleus for further aggregation of oily particles, fecal pellets, phyto- and zooplankton (Mitra et al. 2012), which in combination produced a massive fallout of oil-derived marine snow sinking to the seafloor (Passow et al. 2012; Passow 2016). In addition to particle-associated transport, the hydrocarbon plume in the deep water

(a)
(b)
(c)
(e)
(d)
(f)

◂**Fig. 10 The seafloor near the Macondo wellhead, MC252 and MC118. a** Rough surface texture of the sediment with worm burrows and the tips of dead worms protruding from the sediment. **b** Surface of a sediment core with curled-up dead worms, recovered by multicorer ca. 2 miles west of the Macondo wellhead, and photographed in the *Atlantis* shipboard lab. **c** In situ view of sediment surface littered with small white shells of pteropods (pelagic mollusks), ca. 1 mile west of the Macondo wellhead. **d** Sediment core recovered by multicorer ca. 2 miles south of Macondo wellhead, with 4–5 cm surface layer of red-brown deposits on dark-brown layer of asphaltenes at ca. 5 cm depth, followed by light-colored seafloor sediments underneath. **e** Surviving crabs on shell hash and carbonates at MC118; the crabs were brown-colored instead of red, and unresponsive to the presence of *Alvin*, or its external arm picking up specimens. **f** Filter-feeding gorgonian coral with commensalistic brittle stars, and unresponsive crab, at MC118. Images **a** and **c**: *Alvin* dive 4659, Dec. 2, 2010; Images **b** and **d**: A. Teske, UNC Chapel Hill, **b** collected on Dec. 1 2010, and **d** collected on Nov. 29, 2010; Images **e** and **f**: *Alvin* dive 4658, Dec. 1, 2010

column, containing an estimated 50% of the discharged oil (Joye 2015), impacted slope sediments in the northern Gulf of Mexico (reviewed in Daly et al. 2016). Several lines of geochemical evidence show the arrival of petroleum hydrocarbons on the seafloor. The chemical composition of the sedimented hydrocarbons was congruent with Macondo oil as its source (Mason et al. 2014). Widespread sedimentation of fossil petrocarbon, equivalent to approximately 3–5% of the total carbon emitted during the Deepwater Horizon blowout, was demonstrated by analysis of isotopic composition ($\delta^{13}C$ and $\Delta^{14}C$) for sedimentary carbon in seafloor sediments (Chanton et al. 2015). Sedimentation rates accelerated by two orders of magnitude during summer and early fall of 2010, as determined by analyzing the ^{234}Th inventories of surficial sediment cores (Brooks et al. 2015; Joye et al. 2019). The sedimentation pulse directly impacted redox conditions in the sediment as it was shown that the hydrocarbons and marine snow-derived organic matter catalyzed microbial metal reduction, which led to iron mobilization and subsequent iron reoxidation and precipitation in the surficial sediment layers (Hastings et al. 2016), visible as the conspicuous and persistent reddish-brown color in oil-impacted cores (Fig. 10d).

A microbial census of the oil-impacted MC252 sediments collected during 2010 and 2011 showed that aerobic heterotrophic bacteria such as *Roseobacter*, *Verrucomicrobiaceae*, and *Bacteroidetes* dominated in September and October 2010 (Yang et al. 2016a; Joye et al. 2019), most likely by responding to the availability of freshly deposited oil-derived marine snow. It can be hypothesized that a stimulation of heterotrophic activity leads to localized oxygen depletion and the formation of anoxic microniches at the sediment-seawater interface. These would then provide suitable habitats and substrates for a wide range of anaerobes, such as denitrifying bacteria (Mason et al. 2014), metal-reducing bacteria (Hastings et al. 2016), or sulfate-reducing bacteria (Kimes et al. 2013; Yang et al. 2016a) that continue to degrade recalcitrant petroleum hydrocarbons at the seafloor in situ. The microbial degradation cascade triggered by oil snow deposition on the seafloor seems to have slowed down in November 2010, in the sense that the overall bacterial community structure in the sediment was no longer controlled by the oil sedimentation response, but resembled again the pre-spill community (Yang et al. 2016a). The microbial response to petrocarbon deposition remained detectable on finer taxonomic scales when specialized oil-degrading bacterial populations were tracked, for example, the obligate polycyclic aromatics degrader *Cycloclasticus* that commonly persists in water column and sediment (Kleindienst et al. 2016; Yang et al. 2016a, b). The strong spatial and temporal dynamics of the microbial hydrocarbon response after the Deepwater Horizon blowout invites comparisons to the microbial dynamics of phytoplankton blooms or seasonal anoxia that unfold over time, rather than analogies to mature seep communities that do not experience these fluctuations, as long as their particular seep habitat remains stable.

Following the Deepwater Horizon blowout, the area surrounding the Macondo wellhead was examined in detail for hydrocarbon seeps, the intent being to show that intermittent oil leakage and contamination at MC252 do not originate from an undiscovered large natural oil seep in the vicinity (Conti et al. 2016). Interestingly, this survey uncovered numerous small-scale seep features, such as seafloor carbonates, microbial mats, small hydrate exposures, small mud volcano-like mounds, pockmarks, erosional seafloor depressions, and localized gas seepage. These features were concentrated around the base of salt domes adjacent to the Macondo wellhead, and appear to be linked to faults in the sediment layers as they are fractured by the rising salt dome underneath (Conti et al. 2016).

If we extrapolate from these results that small-scale seep features are widespread and ubiquitous across the entire salt-tectonic seascape on the northern slope of the Gulf of Mexico, an interesting picture emerges. An archipelago of seafloor seeps connects microbial habitats and communities so efficiently that biogeographical isolation of individual locations and particular seep communities is diminished or eliminated, and the local chemistry emerges as the key selection factor that controls microbial composition and activity (Orcutt et al. 2010; Joye et al. 2010; Bowles et al. 2016). By the same token, this high degree of connectivity buffers the microbial communities and associated seep fauna on the northern slope against natural fluctuations in particular seeps (brine lake/mud volcano changes), and against episodic anthropogenic disturbance, with the Deepwater Horizon blowout as the outstanding example. By its extent and habitat variety, the Gulf of Mexico seep landscape provides an unrivaled natural microbial seafloor laboratory to investigate this interplay of underlying unity and local divergence in microbial community structure and activity (see Chap. 8 for a specific microbial example).

References

Amos S (1987) Late Triassic-Jurassic paleogeography and origin of Gulf of Mexico Basin. AAPG Bull 71(4):419–451

Bennett SS, Hanor JS (1987) Dynamics of subsurface salt dissolution at the Welsh Dome, Louisiana Gulf Coast. In: Lerche I, O'Brien JJ (eds) Dynamical geology of salt and related structures. Academic Press, Orlando, pp 653–677

Bowles MW, Joye SB (2011) High rates of denitrification and nitrate removal in cold seep sediments. ISME J 5:565–567

Bowles MW, Samarkin VA, Bowles K, Joye SB (2011) Weak coupling between sulfate reduction and the anaerobic oxidation of methane in methane-rich seafloor sediments during ex-situ incubation. Geochim Cosmochim Acta 75:500–519

Bowles MW, Hunter KS, Samarkin VA, Joye SB (2016) Variations in geochemical signatures and microbial activity within and between diverse cold seep habitats along the lower continental slope of the Gulf of Mexico. Deep-Sea Res II 129:31–40

Bowles MW, Samarkin VA, Hunter KS, Dowell E, Finke N, Teske AP, Girguis PR, Joye SB (2019) Remarkable capacity for anaerobic oxidation of methane at high methane concentration. Geophys Res Lett, in revision

Brooks JM, Kennicutt MC, Fay RR, McDonald TJ, Sassen R (1984) Thermogenic gas hydrates in the Gulf of Mexico. Science 223:696–698

Brooks JM, Cox HB, Bryan WR, Kennicutt MC, Mann RG, MacDonald TJ (1986) Association of gas hydrates and oil seepage in the Gulf of Mexico. Org Geochem 10:221–234

Brooks GR, Larson RA, Flower B, Hollander D, Schwing PT, Robero I, Moore C, Reichart G-J, Jilbert T, Chanton JP, Hastings DW, Overholt WA, Marks KP, Kostka JE, Holmes CW, Hollander D (2015) Sedimentation pulse in the NE Gulf of Mexico following the 2010 DWH blowout. PLoS ONE 10:e0132341

Bryant RW, Bryant JR, Feeley MH, Simmons GR (1990) Physiographic and bathymetric characteristics of the continental slope, Northwest Gulf of Mexico. Geo-Mar Lett 10:182–199

Chanton JP, Zhao T, Rosenheim B, Joye SB, Bosman S, Brunner C, Yaeger K, Hollander D (2015) Radiocarbon tracing of the flux of petrocarbon to the sea floor associated with the Deepwater Horizon event. Environ Sci Technol 49:847–854

Chen SC, Musat N, Lechtenfeld OJ, Paschke H, Schmidt M, Said N, Popp D, Calabrese F, Stryhanyuk H, Jaekel U, Zhu Y-G, Joye SB, Richnow HH, Widdel F, Musat F (2019) Anaerobic oxidation of ethane by archaea from a marine hydrocarbon seep. Nature 568:108–111

Conti A, D'Emidio M, Macelloni L, Lutken C, Asper V, Woolsey M, Jarnagin R (2016) Morpho-acoustic characterization of natural seepage features near the Macondo wellhead (ECOGIG site OC26, Gulf of Mexico). Deep-Sea Res II 129:53–65

Crespo-Medina M, Bowles MW, Samarkin VA, Hunter KS, Joye SB (2016) Microbial diversity and activity in seafloor brine lake sediments (Alaminos Canyon block 601, Gulf of Mexico). Geobiology 14:483–498

D'souza NA, Subramaniam A, Juhl AR, Hafez M, Chekalyuk A, Phan S, Yan B, MacDonald IR, Weber SC, Montoya JP (2016) Elevated surface chlorophyll associated with natural oil seeps in the Gulf of Mexico. Nat Geosci 9:215–218

Daly K, Passow U, Chanton J, Hollander D (2016) Assessing the impacts of oil-associated marine snow formation and sedimentation during and after the Deepwater Horizon oil spill. Anthropocene 13:18–33

Garcia-Pineda O, MacDonald IR, Zimmer B, Shedd B, Roberts H (2010) Remote- sensing evaluation of geophysical anomaly sites in the outer continental slope, northern Gulf of Mexico. Deep-Sea Res II 57:1859–1869

Haeckel M, Suess E, Wallmann K, Rickert D (2004) Rising methane gas bubbles form massive hydrate layers at the seafloor. Geochim Cosmochim Acta 68:4335–4345

Hastings DW, Schwing PT, Brooks GR, Larson RA, Morford JL, Roeder T, Quinn KA, Bartlett T, Romero IC, Hollander DJ (2016) Changes in sediment redox conditions following the BP DWH blowout event. Deep-Sea Res II 129:167–178

Hoshino T, Toki T, Ijiri A, Morono Y, Machiyama H, Ashi J, Okamura K, Inagaki F (2017) Atribacteria from the subseafloor sedimentary biosphere disperse to the hydrosphere through submarine mud volcanoes. Front Microbiol 8:1135

Johansen C, Todd AC, MacDonald IR (2017) Time series video analysis of bubble release processes at natural hydrocarbon seeps in the northern Gulf of Mexico. Mar Pet Geol 82:21–34

Joye SB (2015) Deepwater Horizon, 5 years on. Science 349:592–593

Joye SB, MacDonald IR, Montoya JP, Peccini M (2005) Geophysical and geochemical signatures of Gulf of Mexico seafloor brines. Biogeosciences 2:295–309

Joye SB, Samarkin VA, Orcutt BN, MacDonald IR, Hinrichs K-U, Elvert M, Teske A, Lloyd KG, Lever MA, Montoya JP, Meile CD (2009) Metabolic variability in seafloor brines revealed by carbon and sulphur dynamics. Nat Geosci 2:349–354

Joye SB, Bowles MW, Samarkin VA, Hunter KS, Niemann H (2010) Biogeochemical signatures and microbial activity of different cold seep habitats along the Gulf of Mexico lower slope. Deep-Sea Res II 57:1990–2001

Joye SB, Hunter KS, Medeiros P, Benitez-Nelson C, Moore WS, Asper V, Diercks A, Passow U, Yang T, Teske A, Montoya JP (2019) Soot and slime: Burning and microbial metabolism altered and transported Macondo oil from the sea surface to the seabed. Elementa: Science of the Anthropocene, submitted

Kennicutt MC, Brooks JM, Denoux GJ (1988) Leakage of deep, reservoired petroleum to the near surface on the Gulf of Mexico continental slope. Mar Chem 24:39–59

Kimes NE, Callaghan AV, Aktas DF, Smith WL, Sunner J, Golding B, Drozdowska M, Hazen TC, Suflita JM, Morris PJ (2013) Genomic analysis and metabolite profiling of deep-sea sediments from the Gulf of Mexico following the Deepwater Horizon oil spill. Front Microbiol 4:50

Kleindienst S, Grim S, Sogin M, Bracco A, Crespo-Medina M, Joye SB (2016) Diverse, rare microbial taxa responded to the Deepwater Horizon deep-sea hydrocarbon plume. ISME J 10:400–415

Kniemeyer O, Musat F, Sievert SM, Knittel K, Wilkes H, Blumenberg M, Michaelis W, Classen A, Bolm C, Joye SB, Widdel F (2007) Anaerobic oxidation of short-chain hydrocarbons by marine sulphate-reducing bacteria. Nature 449:898–901

Knittel K, Boetius A, Lemke A, Eilers H, Lochte K, Pfannkuche O, Linke P, Amann R (2003) Activity, distribution, and diversity of sulfate reducers and other bacteria in sediments above gas hydrate (Cascadia Margin, Oregon). Geomicrobiol J 20:269–294

Lapham LL, Chanton JP, Martens CS, Woolsey RJ (2008) Microbial activity in surficial sediments overlying acoustic wipe-out zones at a Gulf of Mexico cold seep. Geochem Geophys Geosyst 9:Q06001

Lapham L, Chanton JP, Chapman R, Martens CS (2010) Methane under-saturated fluids in deep-sea sediments: Implications for gas hydrate stability and rates of dissolution. Earth Planet Sci Lett 298:275–285

Lapham LL, Wilson R, Riedel M, Paull CK, Holmes ME (2013) Temporal variability of in situ methane concentrations in gas-hydrate bearing sediments near Bullseye Vent, Northern Cascadia Margin. Geochem Geophys Geosyst 14:2445–2459

Laso-Pérez R, Wegener G, Knittel K, Widdel F, Harding KJ, Krukenberg V, Meier DV, Richter M, Tegetmeyer HE, Riedel D, Richnow H-H, Adrian L, Reemtsma T, Lechtenfeld O, Musat F (2016) Thermophilic archaea activate butane via alkyl-coenzyme M formation. Nature 539:396–401

Lerche I, Petersen K (1995) Salt and sediment dynamics. CRC Press, Boca Raton, FL

Lever MA, Teske A (2015) Methane-cycling archaeal diversity in hydrothermal sediment investigated by general and group-specific functional gene and 16S rRNA gene PCR primers. Appl Environ Microbiol 81:1426–1441

Lin Y-S, Heuer VB, Goldhammer T, Kellermann MY, Zabel M, Hinrichs K-U (2012) Towards constraining H_2 concentration in subseafloor sediment: a proposal for combined analysis by two distinct approaches. Geochim Cosmochim Acta 77:186–201

Liu Y, Whitman WB (2008) Metabolic, phylogenetic, and ecological diversity of the methanogenic archaea. Ann NY Acad Sci 1125:171–189

Lloyd KG, Lapham L, Teske A (2006) An anaerobic methane-oxidizing community of ANME-1b archaea in hypersaline Gulf of Mexico sediments. Appl Environ Microbiol 72:7218–7230

Lloyd KG, Albert DB, Biddle JF, Chanton JP, Pizarro O, Teske A (2010) Spatial structure and activity of sedimentary microbial communities underlying a *Beggiatoa* spp. mat in a Gulf of Mexico hydrocarbon seep. PLoS ONE 5:e8738

Lloyd KG, Alperin M, Teske A (2011) Environmental evidence for net methane production and oxidation in putative Anaerobic MEthanotrophic (ANME) archaea. Environ Microbiol 13:2548–2564

MacDonald IR, Peccini MB (2009) Distinct activity phases during the recent geological history of a Gulf of Mexico mud volcano. Mar Petrol Geol 26:1824–1830

MacDonald IR, Reilly JF, Guinasso NL et al (1990) Chemosynthetic mussels at a brine-filled pockmark in the northern Gulf of Mexico. Science 248:1096–1099

MacDonald IR, Buthman DB, Sager WW, Peccini MB, Guinasso NL (2000) Pulsed oil discharge from a mud volcano. Geology 28:907–910

Macelloni L, Brunner CA, Caruso S, Lutken CB, D'Emidio M, Lapham LL (2013) Spatial distribution of seafloor bio-geological and geochemical processes as proxies of fluid flow regime and evolution of a carbonate/hydrates mound, Woolsey Mound. Deep-Sea Res I 74:25–38

Mason OU, Scott NM, Gonzalez A, Robbins-Pianka A, Bælum J, Kimbrel J, Bouskill NJ, Prestat E, Borglin S, Joyner DC, Fortney JL, Jurelevicius D, Stringfellow WR, Alvarez-Cohen L, Hazen TC, Knight R, Gilbert JA, Jansson JK (2014) Metagenomics reveals sediment microbial community response to Deepwater Horizon oil spill. ISME J 8:1464–1475

McGee TM (2006) A seafloor observatory to monitor gas hydrates in the Gulf of Mexico. Leading Edge 25:644–647

McHatton SC, Barry JP, Jannasch HW, Nelson DC (1996) High nitrate concentrations in vacuolate, autotrophic marine *Beggiatoa* spp. Appl Environ Microbiol 62:954–958

McNutt MK, Chu S, Lubchenco L, Hunter T, Dreyfus G, Murawski SA, Kennedy DM (2012) Applications of science and engineering to quantify and control the Deepwater Horizon oil spill. Proc Natl Acad Sci USA 109:20222–20228

Mills HJ, Martinez RJ, Story S, Sobecky PA (2004) Identification of members of the metabolically active microbial populations associated with *Beggiatoa* species mat communities from Gulf of Mexico cold-seep sediments. Appl Environ Microbiol 70:5447–5458

Mitra S, Kimmel DG, Snyder J, Scalise K, McGlaughon BD, Roman MR, Jahn GL, Pierson JJ, Brandt SB, Montoya JP, Rosenbauer RJ, Lorenson TD, Wong FL, Campbell PL (2012) Macondo-1 well oil-derived polycyclic aromatic hydrocarbons in mesozooplankton from the northern Gulf of Mexico. Geophys Res Lett 39:1–7

Nigro LM, Hyde AS, MacGregor BJ, Teske A (2016) Phylogeography, salinity adaptations and metabolic potential of the Candidate Division KB1 Bacteria based on a partial single cell genome. Front Microbiol 7:1266

Nikolaus R, Ammerman J, MacDonald IR (2003) Distinct pigmentation and trophic modes in *Beggiatoa* from hydrocarbon seeps in the Gulf of Mexico. Aquatic Microb Ecol 32:85–93

Orcutt B, Boetius A, Elvert M, Samarkin V, Joye SB (2005) Molecular biogeochemistry of sulfate reduction, methanogenesis and oxidation of methane at Gulf of Mexico cold seeps. Geochim Cosmochim Acta 69:4267–4281

Orcutt BN, Joye SB, Kleindienst S, Knittel K, Ramette A, Reitz A, Samarkin V, Treude T, Boetius A (2010) Impact of natural oil and higher hydrocarbons on microbial diversity, distribution, and activity in Gulf of Mexico cold-seep sediments. Deep Sea Res II 57:2008–2021

Otte S, Kuenen JG, Nielsen LP, Pearl HW, Zopfi J, Schulz HN, Teske A, Strothmann B, Gallardo VA, Jørgensen BB (1999) Nitrogen, carbon, and sulphur metabolism in natural *Thioploca* samples. Appl Environ Microbiol 65:3148–3157

Passow U (2016) Formation of rapidly-sinking, oil-associated marine snow. Deep Sea Res II 129:232–240

Passow U, Ziervogel K, Asper V, Diercks A (2012) Marine snow formation in the aftermath of the Deepwater Horizon oil spill in the Gulf of Mexico. Environ Res Lett 7:035031

Reddy CM, Arey JS, Seewald JS, Sylva SP, Lemkau KL, Nelson RK, Carmichael CA, McIntyre CP, Fenwick J, Ventura GT, Van Mooy BAS, Camilli R (2012) Composition and fate of gas and oil released to the water column during the Deepwater Horizon oil spill. Proc Natl Acad Sci USA 109:20229–20234

Roberts HH, Carney RS (1997) Evidence of episodic fluid, gas, and sediment venting on the Northern Gulf of Mexico Slope. Econ Geol 92:863–879

Roberts HH, Aharon P, Carney R, Larkin J, Sassen R (1990) Sea floor responses to hydrocarbon seeps, Louisiana continental slope. Geo-Mar Lett 10:232–243

Roberts HH, McBride RA, Coleman J (1999) Outer shelf and slope geology of the Gulf of Mexico: an overview. In: Kumpf H, Steidinger K, Sherman K (eds) The Gulf of Mexico large marine ecosystems: assessment, sustainability, and management. Blackwell Science, pp 93–112

Roberts HH, Fisher CR, Brooks JM, Bernard B, Carney RS, Cordes E, Shedd W, Hunt J Jr, Joye SB, MacDonald IR, Morrison C (2007) Exploration of the deep Gulf of Mexico slope using DSV *Alvin*: site selection and geologic character. Gulf Coast Assoc Geol Soc Trans 57:647–659

Roberts HH, Feng D, Joye SB (2010a) Cold-seep carbonates of the middle and lower continental slope, northern Gulf of Mexico. Deep-Sea Res II 57:2040–2054

Roberts HH, Shedd W, Hunt J (2010b) Dive site geology: DSV ALVIN (2006) and ROV JASON II (2007) dives to the middle-lower continental slope, northern Gulf of Mexico. Deep Sea Res II 57:1837–1858

Saager WW, Lee CS, Macdonald IR, Schroeder WW (1999) High-frequency near-bottom acoustic reflection signatures of hydrocarbon seeps on the northern Gulf of Mexico continental slope. Geo-Mar Lett 18:267–276

Salman-Carvalho V, Fadeev E, Joye SB, Teske A (2016) How clonal is clonal? Genome plasticity across multicellular segments of a "Candidatus" Marithrix filament from sulfidic, briny seafloor sediments in the Gulf of Mexico. Front Microbiol 7:1173

Sassen R, Roberts H, Aharon R, Larkin J, Chinn E, Carney R (1993) Chemosynthetic bacterial mats at cold hydrocarbon seeps Gulf of Mexico continental slope. Org Geochem 20:77–89

Sassen R, Joye S, Sweet ST, DeFreitas DA, Milkov AV, MacDonald IR (1999) Thermogenic gas hydrates and hydrocarbon gases in complex chemosynthetic communities Gulf of Mexico continental slope. Org Geochem 30:485–497

Schreiber L, Holler T, Knittel K, Meyerdierks A, Amann R (2010) Identification of the dominant sulfate-reducing bacterial partner of anaerobic methanotrophs of the ANME-2 clade. Environ Microbiol 12:2327–2340

Stevens EN, Bailey JV, Flood BE, Jones DS, Gilhooly WP III, Joye SB, Teske A, Mason OU (2015) Barite encrustation of benthic sulfide-oxidizing bacteria at a marine cold seep. Geobiology 13:588–603

Suzuki D, Li Z, Cui X, Zhang C, Katayama A (2014) Reclassification of *Desulfobacterium anilini* as *Desulfatiglans anilini* comb. nov. within *Desulfatiglans* gen. nov., and description of a 4-chlorophenol-degrading sulfate-reducing bacterium, *Desulfatiglans parachlorophenolica* sp. nov. Int J Syst Evol Microbiol 64:3081–3086

Teske A, Salman V (2014) The family *Beggiatoaceae*. In: Rosenberg E, DeLong EF, Thompson F, Lory S, Stackebrandt E (eds) The prokaryotes – gammaproteobacteria. The Prokaryotes, 4th edn. Springer, Berlin/Heidelberg, Chapter 6, pp 93–134. https://www.doi.org/10.1007/978-3-642-38922-1_290

Teske A (2019) Hydrocarbon-degrading microbial communities in natural oil seeps. In: McGenity TJ (ed) Handbook of hydrocarbon and lipid microbiology. Microbial communities utilizing hydrocarbons and lipids: members, metagenomic and ecophysiology. Springer

Torres ME, Wallmann K, Tréhu AM, Bohrmann G, Borowski WS, Tomaru H (2004) Gas hydrate growth, methane transport, and chloride enrichment at the southern summit of Hydrate Ridge, Cascadia margin off Oregon. Earth Planet Sci Lett 226:225–241

Underwood S, Lapham L, Teske A, Lloyd KG (2016) Microbial community structure and methane-cycling activity of subsurface sediments at Mississippi Canyon 118 before the Deepwater Horizon Disaster. Deep-Sea Res II 129:148–156

Van Capellen P, Viollier E, Roychoudhury A, Clark L, Ingall E, Lowe K, Dichristina T (1998) Biogeochemical cycles of manganese and iron at the oxic-anoxic transition of a stratified marine basin (Orca Basin, Gulf of Mexico). Environ Sci Technol 32:2931–2939

Wankel SD, Joye SB, Samarkin VA, Shah SR, Friedrich G, Melas-Kyriazi J, Girguis PR (2010) New constraints on methane fluxes and rates of anaerobic methane oxidation in a Gulf of Mexico brine pool via in situ mass spectrometry. Deep-Sea Res II 57:2022–2029

Widdel F, Rabus R (2001) Anaerobic biodegradation of saturated and aromatic hydrocarbons. Curr Opin Biotechnol 12:259–276

Yang T, Speare K, McKay LJ, MacGregor BJ, Joye SB, Teske A (2016a) Distinct bacterial communities in surficial seafloor sediments following the 2010 Deepwater Horizon blowout. Front Microbiol 7:1384

Yang T, Nigro LM, Gutierrez T, D'Ambrosio L, Joye SB, Highsmith R, Teske A (2016b) Pulsed blooms and persistent oil-degrading bacterial populations in the water column during and after the Deepwater Horizon blowout. Deep-Sea Res II 129:282–291

Zhuang G, Elling F, Nigro LM, Samarkin V, Joye SB, Teske A, Hinrichs K-U (2016) Multiple evidence for methylotrophic methanogenesis as the dominant methanogenic pathway in hypersaline sediments from the Orca Basin, Gulf of Mexico. Geochim Cosmochim Acta 187:1–20

Ziervogel K, McKay LJ, Rhodes B, Osburn CL, Dickson-Brown J, Arnosti C , Teske A (2012) Microbial activities and dissolved organic matter dynamics in oil-contaminated surface seawater from the Deepwater Horizon oil spill site. PLoS One 7(4):e34816

Chapter 5
Benthic Deep-Sea Life Associated with Asphaltic Hydrocarbon Emissions in the Southern Gulf of Mexico

Gunter Wegener, Katrin Knittel, Gerhard Bohrmann and Florence Schubotz

Abstract At the Campeche Knolls in the southern Gulf of Mexico large-scale hydrocarbon emissions are associated with numerous salt tectonic structures. A notable feature of this area is the expulsion of highly viscous heavy oils, also referred to as asphalts, which form lava-like flows on the seafloor. These oil and asphalt seeps have been detected via satellite imaging of oil slicks at the ocean surface and by acoustically-detected gas emission sites in the water column. Here, we describe the type locality 'Chapopote' (Aztec word for tar) where the expelled hydrocarbons provide an energy source for microorganisms and subsequently the deep-sea fauna. In deeper asphalt layers and sediment, 16S rRNA gene sequences of archaea affiliated with *Methanosaeta* and bacteria of the *Syntrophaceae* were detected. Together with the observed light methane isotope values this indicates biogenic methanogenesis from aliphatic hydrocarbons at these depths. In shallow oil-soaked sediments gene sequences of hydrocarbon-degrading sulfate-reducing *Deltaproteobacteria* were found. These sediments showed high sulfate reduction but only minor methane oxidation rates. Asphalts and oil-soaked sediments freshly exposed to the surface were densely covered by microbial mats, signaling high metabolic activity. The surface microbial communities were dominated by diverse *Gammaproteobacteria* that likely oxidized hydrocarbons and reduced sulfur compounds. Surface asphalts were colonized by tube worms, mussels and sponges that host bacterial symbionts. In addition, grazing invertebrates such as sea cucumbers, shrimps, and crabs evidently fed on the microbial mats. This close link between the

G. Wegener (✉) · K. Knittel
Max Planck Institute for Marine Microbiology, Bremen, Germany
e-mail: gwegener@mpi-bremen.de

K. Knittel
e-mail: kknittel@mpi-bremen.de

G. Wegener · G. Bohrmann · F. Schubotz
MARUM, Center for Marine Environmental Sciences and Department of Geosciences of the University of Bremen, Bremen, Germany
e-mail: gbohrmann@marum.de

F. Schubotz
e-mail: fschubotz@marum.de

A. Teske and V. Carvalho (eds.), *Marine Hydrocarbon Seeps*,
Springer Oceanography, https://doi.org/10.1007/978-3-030-34827-4_5

microbial and macro-benthic life demonstrates an efficient energy transfer to higher trophic levels from microbial hydrocarbon oxidation at the base of this unique asphalt-hosted ecosystem.

Introduction

Petroleum hydrocarbons fuel modern societies, however, their exploitation particularly from marine reservoirs is becoming increasingly challenging and harmful to the environment. Man-made oil spills are recurring events, as most devastatingly demonstrated by the catastrophic Deep Water Horizon well blow-out in 2010, which has led to severe damage of coastal and pelagic ecosystems (Lin and Mendelssohn 2012; Rivers et al. 2013). In many hydrocarbon provinces, crude oil is also naturally emitted from the seafloor. Hereby, potentially harmful petroleum substances are cycled naturally, and significantly contribute to the energy supply of benthic communities, feeding light-independent hetero- and autotrophic hydrocarbon-degrading microorganisms (Kvenvolden and Cooper 2003; MacDonald et al. 2002, see also Chaps. 1 and 2). Microbial respiration rates at oil seeps are often strongly elevated compared to the marine background with oxygen consumption rates of up to several 100 mmol per m^2 per day (e.g. Boetius and Wenzhöfer 2013; Sommer et al. 2010). These rates are caused by both, direct oxidation of hydrocarbons, and by the oxidation of sulfide, one of the products of the anaerobic degradation of hydrocarbons with sulfate.

The aerobic degradation of hydrocarbons by microorganisms was discovered more than 100 years ago (Rahn 1906; Söhngen 1906) and is well explored since then. Methane, the smallest hydrocarbon, is aerobically oxidized by different strains of *Beta-* and *Gammaproteobacteria* that rely on molecular oxygen for the activation of their substrate. For this, two specific types of methane monooxygenases are known (reviewed by Hanson and Hanson 1996). The aerobic degradation of non-methane hydrocarbon compounds is widespread in the bacterial domain, and also yeast, fungi, and even some algae are known degraders of *n*-alkanes (Van Beilen et al. 2003). Various strains within *Beta-* and *Gammaproteobacteria*, *Firmicutes*, *Actinobacteria*, and some archaea are degraders of aromatic compounds (Habe and Omori 2003; Olajire and Essien 2014).

Anaerobic hydrocarbon degradation has been investigated more recently, and as these processes often yield only little energy they require an interdependent complex network of microbial communities. For instance, the anaerobic oxidation of methane (AOM) is performed by a consortium of anaerobic methanotrophic archaea (ANME) and sulfate-reducing bacteria (Boetius et al. 2000; Orphan et al. 2002; see also Chap. 1, Microbial communities and metabolisms at hydrocarbon seeps). The anaerobic oxidation of specific *n*-alkanes such as hexadecane and aromatic compounds such as naphthalene has been described for sulfate-reducing *Deltaproteobacteria*, mainly for those of the *Desulfosarcina/Desulfococcus* cluster

(Aeckersberg et al. 1991; Kleindienst et al. 2014; Musat et al. 2009; Widdel and Rabus 2001; see also Chap. 2). However, there is also rising evidence for a role of some archaea in the degradation of short-chain hydrocarbons (Laso-Pérez et al. 2016). In the absence of sulfate, degradation of long-chain hydrocarbons to methane and carbon dioxide can be achieved by consortia of bacteria and archaea (Berdugo-Clavijo and Gieg 2014; Zengler et al. 1999). Due to the different affinities of substrates to biodegradation and the specialization of organisms for specific substrates, the loss of petroleum-derived compounds includes complex interspecies collaborations and dependencies and is evidenced by a sequential removal of specific hydrocarbons (Head et al. 2006). The sulfide that is produced during anaerobic hydrocarbon degradation at oil seep and vent sites may attract an extensive colonization by sulfide-oxidizing bacteria at the sediment-water interface, many of which are mat-forming (Jørgensen 1977; Larkin et al. 1994; see also Chap. 8). In turn, other organisms benefit from the large amounts of biomass produced during sulfide oxidation, including invertebrates that act as grazers or hosts of chemosynthetic bacteria (Dubilier et al. 2008; Van Dover and Fry 1994).

Here, we summarize the geological background of hydrocarbon seepage in the southern Gulf of Mexico, and the specific characteristics at the type locality Chapopote, the first described asphalt volcano (MacDonald et al. 2004). We discuss the chemical composition of the asphalts and accompanying volatile hydrocarbons, and explain how these are available as substrates for specific microorganisms thriving at the distinct asphalt types. This journey ends with a description of chemosynthetic macrofauna benefiting from this unique energy source at 3000 m water depth in the dark ocean.

The Hydrocarbon Province in the Southern Gulf of Mexico

The Campeche Knolls are located in the eastern part of the giant Pimienta-Tamabra petroleum system, bound to the south by the Bay of Campeche and in the east by the Banco de Campeche (Fig. 1). This area is characterized by extensive salt tectonism as evidenced by the presence of numerous knolls and ridges, including the Chapopote Knoll (Bryant et al. 1991; Ding et al. 2008; Watkins et al. 1978). Most of the salt was originally deposited during the opening of the Gulf in the Late Jurassic. Within the Jurassic and early Cretaceous, thick layers of organic-rich shales were deposited creating prolific source rocks for the formation of crude oil and natural gas. These horizons are overlain by Cretaceous carbonate reservoir rocks. Increasing sedimentation in the Neogene caused salt tectonic movement (Salvador 1991) creating a network of fractures and structural traps in overlying source and reservoir rocks. These act as conduits and barriers for upwardly migrating oil and gas (Magoon et al. 2001).

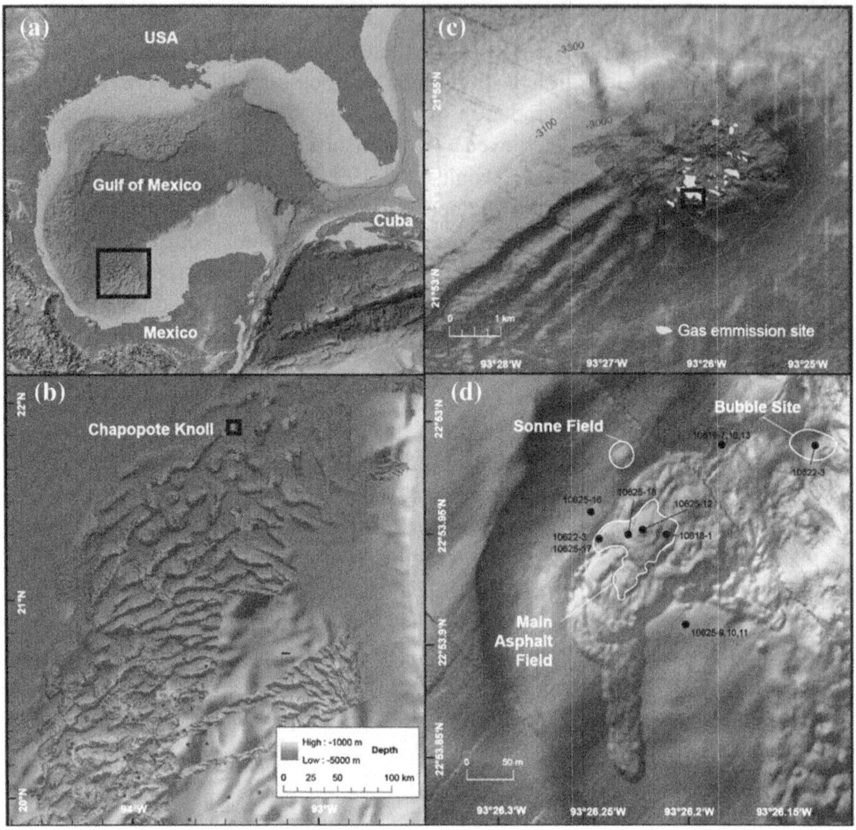

Fig. 1 Geomorphology of the Campeche Knolls and the Chapopote Asphalt Volcano. **a** Location of the Campeche Knolls (black box) within the Gulf of Mexico. **b** Bathymetric map of the Campeche Knolls and oil slick locations (green and grey dots, after Sahling et al. 2016) **c** Bathymetry of the Chapopote Asphalt Volcano, overlain by AUV-derived micro-bathymetry at the crater area with locations of detected gas emission sites. The black box shows the area of the main asphalt field outlined in map D. **d** Micro-bathymetry of the main asphalt area (data from Marcon et al. 2018) and sample sites (for details see Table 1)

Natural petroleum seepage is a recurring feature in the hydrocarbon provinces of the Gulf of Mexico and results in the formation of oil slicks that are visible from space via satellite imaging (MacDonald et al. 1996; see also Chap. 4). Guided by known locations of oil slicks in the southern Gulf, ship-based surveys could demonstrate active hydrocarbon seepage at more than 40 knolls in the Campeche area. Using hydro acoustic echo sounder techniques, many gas emission sites at the seafloor and in the water column were sensed (Sahling et al. 2016). Many of the detected oil and gas emission sites have been investigated further subsequently using towed camera systems, gravity coring, and remotely operated submarines (Brüning et al. 2010; MacDonald et al. 2004; Sahling et al. 2016; Marcon et al.

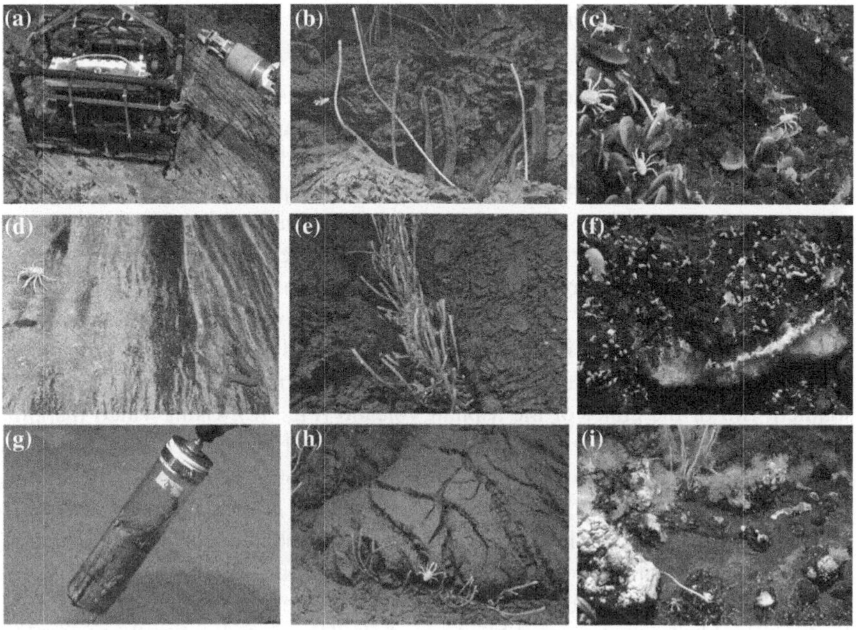

Fig. 2 Seafloor observations of the Chapopote asphalt field using the ROV QUEST during RV METEOR Expedition M67/2 (2006). For further details and seafloor impressions see Brüning et al. 2010; Sahling et al. 2016; Marcon et al. 2018. **a** Benthic chamber deployment on bacterial mat-covered asphalts at the main asphalt field. **b** Brittle and fractured asphalt covered by sediment at the main asphalt field. **c** Asphalt at the "Bubble Site" colonized by mussels, bivalves, crabs, shrimps, and sponges. **d** Grazing sea cucumber and crab on top of microbial mat-covered asphalt at the main asphalt field. **e** Tubeworms growing in between two asphalt flows at the main asphalt field. **f** White gas hydrate grown under dark asphalt deposits, covered by bivalves and snails at the "Bubble Site". **g** Oily sediments sampled by push coring next to the asphalt beds. **h** Fractured asphalt flow at main asphalt field. **i** Weathered asphalt as substrate for sea anemones, sponges, corals, and bacterial mats

2018). Photographic and bathymetric evidence demonstrate that many of the salt features of the Campeche Knolls are covered by expulsed heavy oil that spreads on the seafloor in lava-like flows of solidified asphalt in depths down to 3100 m below sea level. These sites also host chemosymbiotic fauna, comparable to petroleum seeps in the northern Gulf of Mexico (see Chap. 4), but visibly adapted to life on heavy oil, as tubeworms and mussels root within the asphalt layers, and shrimps, sea cucumbers and crabs notably feed of microbial mats that are covering the asphalts (Fig. 2). Other notable features of this unique deepwater asphalt seep system include extensive gas hydrate deposits and oil-laden anoxic sediments locally covered by bacterial mats (Sahling et al. 2016).

The Chapopote Asphalt Volcano

Although asphalt deposits are known from observations and sampling at seven sites in the Campeche Knoll area (Sahling et al. 2016), Chapopote is the location where most extensive asphalt-related investigations were conducted. Chapopote represents a conic elevation of 350–800 m above the abyssal plain (Brüning et al. 2010), which is the result of the underlying salt diapir (Ding et al. 2008). The main structure is ~3 km in diameter with a crater-like depression in its center. The term asphalt volcanism was introduced at Chapopote, as a novel type of hydrocarbon seepage by MacDonald et al. (2004) after an extensive area of solidified asphalt resembling lava-flows in visual appearance was found on top of the knoll.

A prominent asphalt field covering more than 2600 m^2 has been observed on the southern rim of the crater of the asphalt volcano (Fig. 1c). Using video mapping, coring, and hydro-acoustic mapping more widespread but patchy asphalt deposits and gas flares have been confirmed within the crater structure, as well as all along the rim (Brüning et al. 2010; Sahling et al. 2016). A recently published photo mosaic of nearly 1000 single images from the main asphalt field (Fig. 1d) displays a large coverage, and provides detailed site survey information (Marcon et al. 2018). Three continuous flow units construct this field, whereas the oldest flow unit (Unit 1) shows highly fragmented and blocky asphalt. The second oldest flow unit (Unit 2) is characterized by fragmented, deeply fractured asphalt, in which ropy flow shapes remain visible. The youngest, least-altered asphalt deposits (Unit 3) form ropy to very smooth morphologies, and are covered by microbial mats with various density. The densest mats were concentrated near the center of the main asphalt field, where asphalt deposits are the freshest. The density of microbial mats decreased gradually from the core towards the edges of the deposits of the third flow unit (Unit 3), where white mats are only visible within grooves and cracks of the asphalt (Marcon et al. 2018). The authors interpret the flow units to represent at least three major eruption phases, whereas the time scale of an eruption phase and the quiescent time between eruption phases are not clear. The AUV-derived micro-bathymetry data from the area (Fig. 1d) indicate that the asphalt units mapped in the mosaic apparently overlie some additional flow features, which are interpreted as older asphalt deposits (Marcon et al. 2018). The investigations show that the asphalt volcanism is not a continuous process, in which viscous asphalt would slowly seep out of the seabed. Instead, asphalt erupts for a relatively short phase, separated by potentially longer intervals of no eruption activity. The seismic data from Ding et al. (2008) suggested the Chapopote asphalt originates from isolated shallow reservoirs located between 100 and 200 m sediment depth, and that geological events may trigger the asphalt eruptions.

Although the mechanisms causing asphalt seepage and deposition are not yet fully understood, the present assumption is that gas-rich heavy oil is being transported upwards along fractures that are most probably created by salt movements (Marcon et al. 2018). Ductile asphalt, presumably representing more recently emitted material, still contains high amounts of methane (more than 150 mmol per

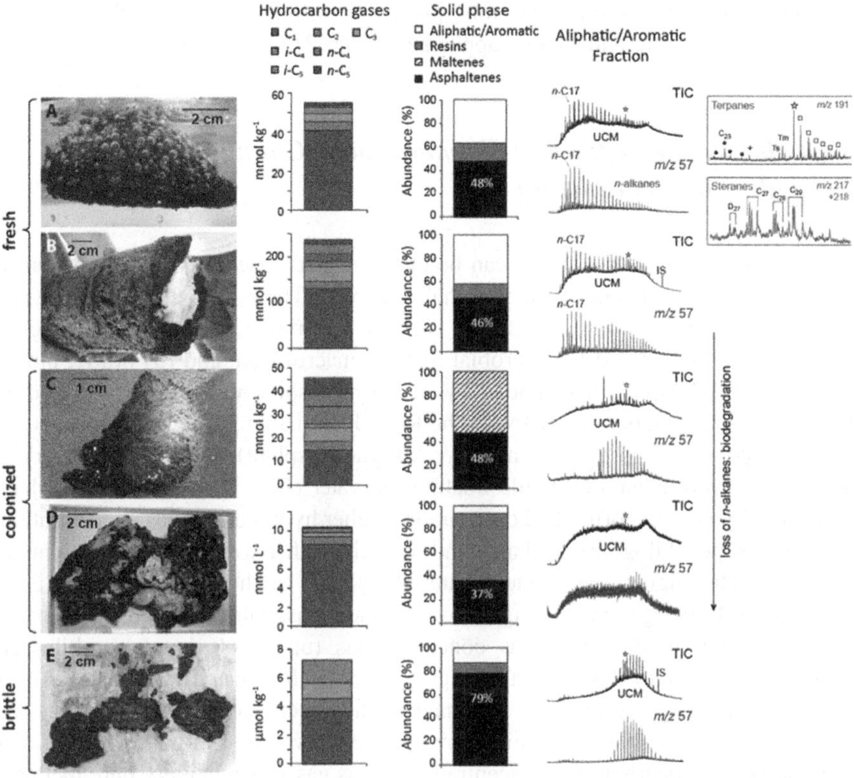

Fig. 3 Representative samples studied for organic matter composition. Shown are hydrocarbon gas distributions, asphaltene, resins and aliphatic/aromatic hydrocarbon contents of the asphalt samples. Total ion chromatograms (TIC) and extracted mass chromatogram *m/z* 57 of the aliphatic/aromatic fraction all show a large unresolved complex mixture (UCM) and differences in *n*-alkane distribution. Representative extracted mass chromatograms of terpanes (*m/z* 191) and steranes (*m/z* 217+218) are shown, which were similar for all analyzed asphalts. **a** GeoB10625-16, bubbling asphalt with white precipitate, **b** GeoB10618–1, deep asphalt with gas hydrate, **c** GeoB10617-6 at 1 cm depth, surface asphalt covered by microbial mat, **d** GeoB10622-3, asphalt covered with sponge, **e** GeoB10619-17, brittle asphalt covered with sediment. Abbreviations: n-C_{17}, C_{17} *n*-alkane; IS, injection standard. Hydrocarbon gas samples were immediately collected in gas tight vials and analyzed by gas chromatography coupled to flame ionization detection. Organic phases were solvent extracted and separated into different fractions as described by Schubotz et al. (2011a). Aliphatic/aromatic hydrocarbon fractions were analyzed by gas chromatography coupled to mass spectrometry. For details see text

kilogram asphalt; Schubotz et al. 2011a) and visible pockets of gas hydrate (Fig. 3), which may facilitate upward migration of the heavy oil. Within the sediment, the oil exhibits positive buoyancy, but after contact with seawater the lighter hydrocarbon gases and other light-end oil components of the seeped oil and asphalt are readily lost due to biological oxidation and abiotic chemical dissolution (water washing), resulting in the negative buoyancy of the settled asphalt flows and patches on the seafloor. Over time, abiotic and biotic processes further lead to asphalt fracturing, resulting in its brittle and blocky appearance (Fig. 2).

Asphalt Geochemistry and Compositional Changes Related to Biodegradation

Distinct asphalt alteration stages can be distinguished based on visual appearance, texture and gas content of the asphalt deposit. There are three main types: (1) seemingly freshly emitted ductile asphalt material, (2) rougher and slightly altered asphalt colonized by microbial mats or macrofauna, and (3) brittle, blocky and sedimented asphalt surrounded by fragmented shell material (Figs. 2 and 3). Ductile, fresh asphalts collected with the ROV had visible gas hydrate pockets and high gas contents (up to 55 mmol per kilogram asphalt) that upon recovery to atmosphere were degassing while floating on water (Fig. 3a). Subsurface asphalts recovered by gravity coring had comparably higher hydrocarbon gas concentrations (200 mmol per kilogram asphalt) and contained fist-size gas hydrate chunks (Fig. 3b). The higher gas contents are likely due to the fact that these cores experienced less degassing during recovery. Ductile asphalt colonized by microbial mats had high hydrocarbon gas concentrations (up to 60 mmol per kilogram asphalt), but contained less methane compared to the other two "fresh" asphalt samples, indicating that methane may have already been partly consumed (Figs. 2d, and 3c). Blockier asphalt, colonized by sponges, bivalves and clams had comparably lower hydrocarbon gas concentrations. This gas was relatively pure methane, possibly because it derives from gas hydrates that were found admixed to the asphalt (Figs. 2f, and 3d). The lowest gas content (8 μmol per kilogram asphalt) was observed in brittle and fragmented asphalt pieces that were already largely covered by sediment, indicating older material where most of the hydrocarbon gases have already been degassed (Fig. 2b, and 3e). The absolute gas concentrations measured in the samples are strongly influenced by degassing during asphalt recovery, but all gas samples were placed in gas tight vials immediately upon recovery of the asphalt samples to reduce gas loss. All asphalts contained abundant methane, but also all other hydrocarbon gases and their isomers (C_1 to C_6) were present, with C_1/C_2 ratios ranging between 4 and 10 pointing to mainly thermogenic sources of the methane (Bernard et al. 1976). A predominantly thermogenic methane source is supported by stable carbon isotope composition ($\delta^{13}C$) of methane with values of approximately −40‰ in all asphalt samples. Only gas hydrate methane and methane of oil-rich sediments exhibit more depleted $\delta^{13}C$-values of −55‰ to −65‰, respectively, indicating biogenic methanogenesis from hydrocarbons in those layers (Fig. 4, Schubotz et al. 2011a).

To further classify the alteration of the different asphalts geochemically their solubility in organic solvents can be determined. Components insoluble in *n*-hexane are referred to as asphaltenes, and represent a complex mixture of high molecular weight macromolecules. All asphalt samples contained high amounts of

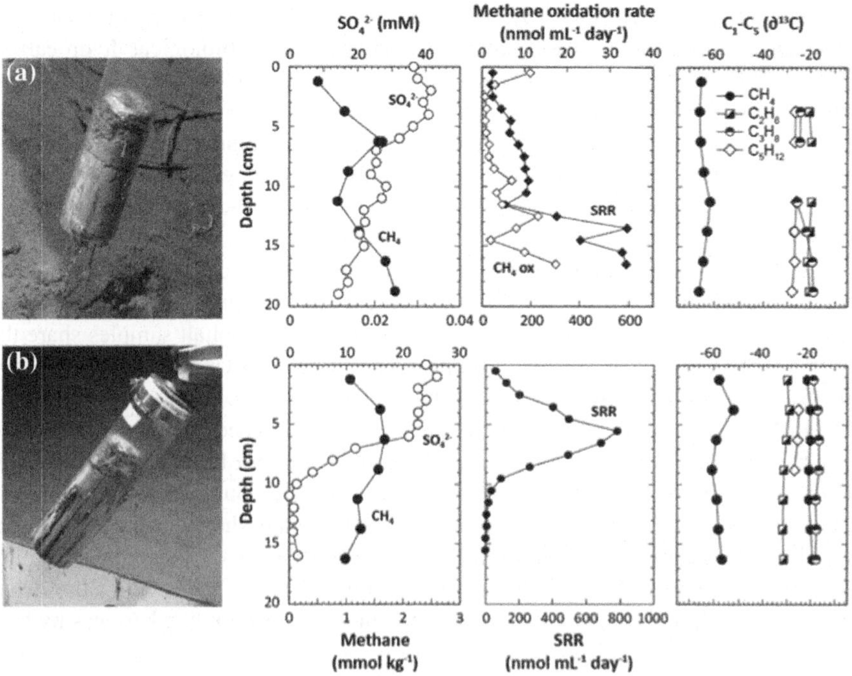

Fig. 4 Biogeochemistry of oil-soaked sediments at Chapopote. Sediments were recovered from two stations, **a** GeoB10619-7, -10, -13 and **b** 10625-9, -10, -11, close to the asphalt fields. The left panel depicts downcore concentrations of methane and sulfate, the middle panel methane oxidation rates and sulfate reduction rates throughout the core, and the right panel shows changes in the stable carbon isotopic composition of hydrocarbon gases methane (CH_4), ethane (C_2H_6), propane (C_3H_8) and pentane (C_5H_{12}). Sulfate concentrations were determined by high-pressure, non-suppressed ion chromatography. Concentrations and stable carbon isotopic composition of hydrocarbon gases were measured via gas chromatography coupled to flame ionization detection and isotope ratio mass spectrometry, respectively (Schubotz et al. 2011a). Rates of methane oxidation and sulfate reduction were determined by ^{14}C-methane and ^{35}S-sulfate tracer incubations, and analytical procedures and calculations were performed as described by Treude et al. (2003) and Kallmeyer et al. (2004)

asphaltenes, accounting for 37–79 weight-% of the sample (Fig. 3). This fraction was not further characterized since an evaluation of molecular structures of asphaltenes requires a combination of elemental, spectroscopic, or ultra-high resolution mass spectral analysis and modeling (e.g., Coelho et al. 2006; Speight 1999). Furthermore, this fraction is generally not considered to be bioavailable to microorganisms. The *n*-hexane soluble fraction, called maltenes, can be further divided into aliphatic and aromatic hydrocarbons, that typically comprise most crude oils, and resins, which contain functionalized and more polar hydrocarbons. The more ductile, non-colonized, fresh asphalts contain about 40% of aliphatic and aromatic hydrocarbons with a resin content of around 15% (Fig. 3). The amount of resins increased to 60% in the sponge-covered asphalt, while the aliphatic/aromatic fraction declined to less than 10%. We primarily assign the loss of the aliphatic and aromatic compounds in the sponge-covered asphalt to biological hydrocarbon degradation and the increase of resins to oxygenation of hydrocarbons during their microbial activation. The exceptionally high asphaltene contents of 79% in the brittle and sedimented asphalt are likely caused by a complete loss of most of the maltene fraction due to biodegradation combined with chemical and physical processes such as water washing (dissolution of hydrocarbons in water, Schubotz 2009).

Further indication for microbial attack of the asphalts can be derived from a more detailed analysis of the aliphatic/aromatic fraction. Biomarker signals (hopane trace *m/z* 191 and sterane trace *m/z* 217) indicate that all asphalt samples share the same source, and match oils that have been described from Upper Thitonian source rocks in the southern Gulf of Mexico (Guzman and Mello 1999; Schubotz et al. 2011a). All studied aliphatic/aromatic fractions are dominated by a large unresolved complex mixture (UCM), which is usually an indication for biodegradation (Fig. 3, Peters et al. 2005). However, the presence of similarly abundant *n*-alkanes in the range from C_{14} to C_{40} in the non-colonized fresh asphalts shows that these samples are little biodegraded (Head et al. 2006; Hostettler and Kvenvolden 1994). Microbial activity is likely inhibited by the limited pore space within the dense asphalts, which prevents circulation of water and nutrients. Only where gas hydrate pockets where observed in the asphalts (Fig. 3b) microbial life may be feasible. Notably, in these samples the *n*-alkanes show a bimodal distribution (higher concentrations of short chain and long chain *n*-alkanes than compounds around C_{20} chain lengths), which can be explained by preferential microbial degradation of short chain *n*-alkanes, resulting in the relative enrichment of the long chain *n*-alkanes (Schubotz et al. 2011a). A complete loss of the short chain (C_{14}–C_{20}) *n*-alkanes was observed in the mat-covered asphalts, indicating that these samples had been moderately biodegraded, following the dogmatic preferential biodegradation of the short chain to long chain *n*-alkanes. The sponge-covered asphalts showed signs of heavy biodegradation, and nearly all *n*-alkanes and low molecular weight aromatic hydrocarbons were depleted (cf. Schubotz 2009; Fig. 3).

Asphalt-Derived Hydrocarbons Induce Increased Microbial Activity

The observed coverage of bacterial mats on the more recently emitted asphalts strongly suggests elevated microbial activity associated to this environment. Indeed, based on benthic chamber measurements, bacterial mat-covered asphalts (station 10625-12) consumed 65–80 mmol oxygen per m^2 per day. In comparison, the non-hydrocarbon influenced pelagic seafloor in the Gulf of Mexico supports an oxygen consumption of approx. $\sim$3 mmol per m^2 (Wenzhöfer and Glud 2002). Hence, the fresh asphalts support a 25 times elevated oxygen consumption rate compared to background sediments. Notably, it is unknown if oxygen consumption is mainly coupled to the aerobic oxidation of hydrocarbons, or, to the oxidation of sulfide, which is suggested by the presence of filamentous, potentially sulfur-oxidizing bacteria (see also Chap. 8).

Also, sediment in the vicinity of asphalt flows was partly covered with microbial mats. Sediments cored from these sites were soaked with heavy oil, and were rich in hydrocarbon gases at concentrations between 0.01 and 1.5 mmol per kilogram sediment (Schubotz et al. 2011a). Sulfate concentrations strongly decreased with depth, resulting in sulfate depletion at 12 cm sediment depth. The rapid sulfate consumption could be linked to intense sulfate reduction with rates up to 800 nmol per cm^3 per day. These rates accumulate to an areal sulfate reduction of 40 mmol per m^2 per day. Due to the heterogeneous nature of the Chapopote asphalt field the measured rates for aerobic and anaerobic processes cannot be truly compared. However, since the nearby measured oxygen consumption of up to 80 mmol per m^2 per day on the likewise mat-covered asphalt may also be predominantly coupled the oxidation of sulfide it is possible that hydrocarbon oxidation mainly proceeds anaerobically, and oxygen consumption is largely coupled to the reoxidation of the sulfide produced during anaerobic hydrocarbon oxidation. Although the potential AOM zone was not fully covered by our measurements, methane oxidation rates in these cores were remarkably low, accounting for only 1% of the concurrently measured sulfate reduction rates in the upper 17 cm (integrated AOM only 0.4 mmol per m^2 per day; Fig. 4). This shows that sulfate reduction was mainly coupled to the oxidation of non-methane hydrocarbons by the microorganisms inhabiting the oily sediments. This observation was supported by fatty acid and archaeal lipid isotope data, showing that AOM was occurring in the sediments, but that the bulk of the bacterial community was rather involved in the degradation of the oil than in the oxidation of isotopically light methane (Schubotz et al. 2011b). At reference sites without visible signs of hydrocarbon seepage, methane concentrations were below 1 μmol per kilogram sediment, AOM was not detectable, and sulfate reduction was low (integrated rates of 1.2 mmol per m^2 per day for the upper 15 cm). In summary, these measurements show that in an otherwise oligotrophic deep ocean environment the asphalt volcanoes provide large amounts of bioavailable hydrocarbons and cause strongly elevated microbial activities.

Microorganisms Associated to Asphalt Volcanism

Different habitats of the Chapopote Asphalt Volcano were sampled to determine their dominant bacterial and archaeal community members (Fig. 5a, b). The asphalt deposits from the main asphalt site were sampled by gravity coring, and liquid or solid asphalts were analyzed independently, representing anoxic, little altered, fresh material. Furthermore, a sediment-asphalt mix was recovered representing the community of the anoxic, but altered and biodegraded material, and the mat-covered surface asphalt was analyzed to search for aerobic community members. The 16S

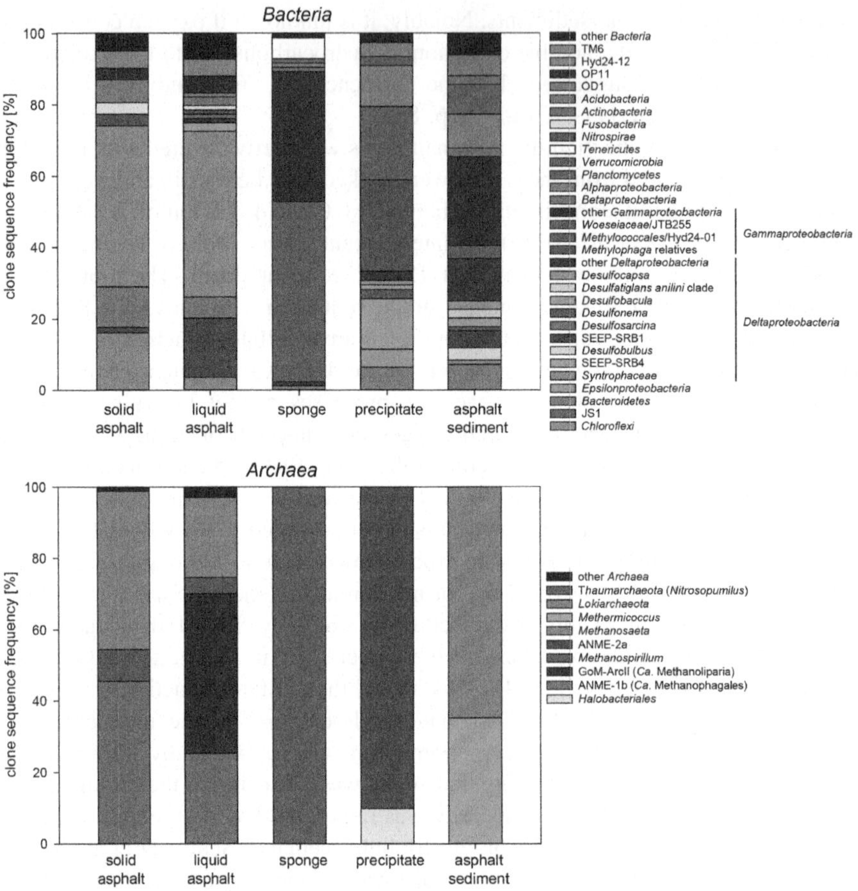

Fig. 5 Relative abundance of bacterial and archaeal 16S rRNA gene sequences retrieved from different hydrocarbon-influenced environments of the Chapopote Asphalt Volcano. DNA was manually extracted according to Zhou et al. (1996) with some modifications for DNA purification from the asphalts. Bacterial and archaeal 16S rRNA genes were amplified using domain-specific primers: GM3F (Muyzer et al. 1995) and EUB1492R (Kane et al. 1993) for *Bacteria*, and ARCH20F-N (modified from Massana et al. 1997; TTC VGG TTG ATC CYG CSR G) and Arc1492R (Teske et al. 2002) for *Archaea*. DNA was ligated in pGEM-T-Easy and transformed into *E. coli* before sequencing. Sequence data was analyzed with the software package ARB (Ludwig et al. 2004) with the SILVA database SSU Ref NR 99, release 123_1 (March 2016). Sequence data reported here are available in the EMBL, GenBank, and DDBJ nucleotide sequence databases under the accession numbers MH885571-MH885636, LC424843-LC424925, LC431881-LC432081

rRNA gene libraries of the subsurface asphalt samples (liquid and solid asphalt) were characterized by very low microbial diversity: about 45% of all bacterial sequences were close relatives to *Syntrophus*, and up to half of the archaeal sequences affiliated with *Methanosaeta*. The co-occurrence of these two clades was previously reported in hydrocarbon-polluted sediments (e.g. Dojka et al. 1998; Watanabe 2001).

Together, organisms of these groups might be able to syntrophically metabolize aliphatic hydrocarbons such as hexadecane, forming methane and carbon dioxide as products (Zengler et al. 1999). The presence of *n*-alkane degraders in this sample corroborates the geochemical findings of a sequential removal of short chain *n*-alkanes described for the gas hydrate-laden subsurface asphalts (see section "Asphalt Geochemistry and Compositional Changes Related to Biodegradation" and Fig. 3b). Interestingly, the asphalts also contained numerous sequences of ANME-1 archaea (46% in solid, and, 25% in liquid asphalts), however, it remains unclear whether either methanogenesis or AOM was a prevailing process in these asphalts. Furthermore, in the liquid asphalt a large number of archaeal sequences aligned with clade GoM-ArcII (*Ca.* Methanoliparia). These archaea have been found repeatedly in seep sediments from the Gulf of Mexico (Orcutt et al. 2010). Based on metagenome assembled genomic information and microscopy *Ca.* Methanoliparia might degrade alkanes and couple this process to methanogenesis without the need of partner organisms (Borrel et al. 2019; Laso-Pérez et al. 2019).

The asphalt sediment mix was inhabited by a highly diverse microbial community comprising all clades of bacteria typically found in marine seep sediments (Knittel et al. 2003; Orphan et al. 2001; Teske et al. 1998) including sequences of different *Deltaproteobacteria.* As most of them are affiliated with potential hydrocarbon-degrading sulfate reducers this data set supports the geochemical evidence for the microbial oxidation of asphalt constituents, such as aliphatic, cyclic, and aromatic compounds under anoxic conditions. Besides anaerobic organisms the asphalt mix contained a large variety of potentially aerobic hydrocarbon degraders such as *Cycloclasticus,* which seem to play a key role in the degradation of hydrocarbons in marine environments (Gutierrez et al. 2013; Kasai et al. 2002). *Cycloclasticus* is also suggested to be a potentially detoxifying symbiont in chemosynthetic mussels and sponges (Rubin-Blum et al. 2017). In contrast to the diverse bacterial community, only two archaeal clades were retrieved from the asphalt sediment mix, *Lokiarchaeota* and *Methermicoccus.* The absence of ANME archaea supports an inferior role of anaerobic oxidation of methane in the asphalts as shown for the oil-rich sediments above (Schubotz et al. 2011b).

Some asphalt surfaces contain sulfur precipitates, but are likely fully oxic, because sequences of organisms with typically aerobic lifestyles were retrieved. These included gammaproteobacterial *Methylophaga*, and *Methylococcales* of the clade Hyd24-01, indicating the importance of aerobic methane oxidation on the surface of the asphalts. A major part of retrieved sequences also affiliated with *Deltaproteobacteria* of the clade SEEP-SRB4 (14% of total sequences), a cluster of yet uncultivated sulfate-reducing bacteria that were abundantly found in surface seep sediments (e.g. Kleindienst et al. 2012; Knittel et al. 2003; Lösekann et al. 2007; Wegener et al. 2008; see also Chap. 2). The presence of these anaerobic microorganisms indicated the rapid oxygen depletion within the asphalts. In addition, large numbers of ANME-2a sequences in the archaeal clone library of the precipitate suggest the presence of AOM. Non-methane alkanes can be oxidized by various organisms, including the aerobic *Gammaproteobacteria*, members of *Bacteroidetes*, and sulfate reducers from the *Deltaproteobacteria* (as partners of

ANME archaea). The large diversity of potential hydrocarbon sources might support a likewise highly diverse microbial community. To describe the microbial communities and their function in greater depths 16S rRNA amplicon sequencing and metagenomic analysis are required.

Chemosynthetic and Grazing Macrofauna

The surface of the asphalt beds is widely colonized by benthic macrofauna. The sessile fauna consists of *Escarpia* sp. tubeworms, diverse sponges, and mussels of the genus *Bathymodiolus*. Furthermore, gastropods such as the galatheid crab *Munidopsis* spp. and the non-pigmented caridean shrimp *Alvinocaris muricola* are particularly abundant in areas of gas seepage (Gaytán-Caballero and Escobar-Briones, personal communication). Together, these organisms display a rich fauna thriving on energy released by oil-degrading microorganisms. Hydrozoans and anemones complete the ecosystem by using the older, brittle asphalt flows as hardground for settlements.

The tubeworms identified as *Escarpia* sp. (Raggi et al. 2013) were concentrated in cracks in or at the boarder of asphalt flows (Fig. 2), suggesting that the worms cannot penetrate the solid flows. Rather, they benefit from sulfide produced during anaerobic hydrocarbon degradation, which is oxidized by their thiotrophic gammaproteobacterial symbionts (Nelson and Fisher 2000; Raggi et al. 2013). Hence the tubeworms and their symbionts indirectly benefit from the asphalts by the activity of sulfate-reducing communities within and below the asphalts. In reverse, the sulfate reducers may benefit from the sulfate-rich water introduced into the asphalts by the activity of the tubeworms.

The mussels *Bathymodiolus brooksi* and *B. heckerae* are typically concentrated on sites showing active oil and gas ebullition (Fig. 2). Both species were shown to host sulfur and methane-oxidizing symbionts (Duperron et al. 2007). Mussels of the group *B. heckerae* analyzed from Chapopote also host the bacteria of the genus *Cycloclasticus*, which are assumed to degrade short-chain hydrocarbons such as butane and propane (Rubin-Blum et al. 2017). Flat and branching sponges were widespread on the asphalt. Sponges host a diverse community of different microorganisms, and are thus likely to benefit from the asphalt-based energy in different ways. A flat sponge tissue collected from the asphalt (sampling location and description Table 1; Fig. 1) was identified as *Poecilosclerida* sp. Cloning of the amplified mitochondrial cytochrome *c* oxidase subunit I gene (COI; for primers see Folmer et al. 1994) showed 99–100% amino acid identity with COI retrieved from sponges found at sites Chapopote and Mictlan of Campeche Knolls (accession numbers MH899018; this study); KU659136 & KU659137; Rubin-Blum et al. 2017). The bacterial 16S rRNA gene sequence library of this tissue contained many sequences of the *Methylococcales*-affiliated uncultivated clade Hyd24-01 (43 of 85 bacterial sequences). This order mainly comprises obligate aerobic methanotrophs (Bowman 2006). Furthermore, the sponge library contained 25 gammaproteobacterial sequences of clade C01-9C-26. Closely related sequences within this cluster

Table 1 Summary of samples and analyses reported in this study

Sample	GeoB-No.	Gear	Position	Description	Analysis
Subsurface asphalts	10618-1	GC	21°53.95′N 93°26.21′W	Solid to liquid asphalt with large gas hydrate chunks, very little sediment	Oil and gas geochemistry, 16S RNA gene library (bacteria/archaea) of liquid and solid asphalt
Oily sediment	10619-7,10,13	PC	21°53.99′N 93°26.18′W	Oil-rich (up to 50%) sediments, partially covered with mats	Porewater geochemistry, sulfate reduction and methane oxidation rates
Oily sediment	10625-9, 10,11	PC	21°53.91′N 93°26.20′W	Oil-rich (up to 50%) sediments, partially covered with mats	Porewater geochemistry, sulfate reduction rates
Brittle asphalt	10619-17	ROV	21°53.95′N 93°26.23′W	Brittle and porous asphalt pieces, covered with sediment	Oil and gas geochemistry
Asphalt with sponge tissue	10622-3	ROV	21°53.99′N 93°26.13′W	Sponge, mussel and bivalve-covered asphalt, collected closely to gas hydrates	Oil and gas geochemistry, 16S rRNA gene library (bacteria/archaea)
Mat-covered asphalt	10625-12	BC	21°53.95′N 93°26.25′W	Fresh asphalt covered with white filamentous bacteria, very little sediment	Oxygen consumption measurements with benthic chamber
Asphalt sediment	10617-6	Rotary bottle	21°53.95′N 93°26.23′W	Solid asphalt pieces mixed with sediment	Oil and gas geochemistry, 16S rRNA gene library (bacteria/archaea)
Asphalt with white precipitate	10625-16	RS	21°53.96′N 93°26.25′W	Solid asphalt with gas hydrates, bubbling and floating on top of water upon recovery, very little sediment	Oil and gas geochemistry, 16S rRNA gene library (bacteria/archaea)
Holothurian	10625-17	SG	21°53.95′N 93°26.25′W	Specimen of *Chiridota heheva* collected from the mat-covered asphalt	Lipid biomarker and stable carbon isotope analyses

Abbreviations *BC* Benthic chamber; *ROV* arm, sampling with the *ROV* manipulator into the biobox, *RS* Samples retrieved by rotary vacuum pump; *GC* Sampling with gravity corer; *SG* slurp gun were retrieved from other marine sponges such as *Ianthella basta* (Luter et al. 2010) and *Inflatella pellicula* (e.g. KF597133), suggesting that these bacteria might have a symbiotic lifestyle. Additionally, sequences of the hydrocarbon degrader *Cycloclasticus* were detected in the sponge, supporting recent observations of short-chain hydrocarbon degraders in this host (Arellano et al. 2013; Rubin-Blum et al. 2017). Furthermore, all archaeal sequences from the sponge tissue belonged to "*Candidatus* Nitrosopumilus", which oxidizes the metabolite ammonium as planktonic organism, or, as symbiont in sponges (Feng et al. 2016; Hallam et al. 2006; Hoffmann et al. 2009; Könneke et al. 2005; Pape et al. 2006).

Non-pigmented crabs and holothurians (sea cucumbers) were abundant grazers in this system. Feeding traces visible on the asphalt mats suggest a preference of this grazer for this habitat (Fig. 2d). To further study the dietary behavior of the dominant holuthoria *Chiridota heheva* two individuals were collected in the vicinity of microbial mats and were analyzed for their gut contents (Fig. 6). The guts of both individuals contained notable quantities of asphalt hydrocarbons (40–60 milligram per gram soft tissue) that resembled the compositions of surficial asphalt (Fig. 3, Schubotz et al. 2011a). These hydrocarbons, however, cannot be metabolized by invertebrates, and are rather ingested accidently by the organisms while they are feeding on the microbial mats. In addition to asphalt pieces, the guts contained large amounts of free fatty acids and other lipids that can be associated to distinct mat community members, including diagnostic markers for AOM partner bacteria such as $C_{16:1\omega5}$, iso- and anteiso $C_{15:0}$ fatty acids, and C_{15}–C_{17} monoalkyl glycerol ethers (MAGE) with depleted $\delta^{13}C$ values around −45‰ (Elvert et al. 2003; Rütters et al. 2001). The remaining C_{14}, C_{16} and C_{18} saturated and mono-unsaturated free fatty acids may be derived from sulfide-oxidizers and other mat-forming bacteria (Kerger et al. 1986). Also diagnostic markers for both aerobic and anaerobic oxidation of methane were detected: 4-methyl-sterols and diplopterol with low stable carbon isotopic composition of −65‰ to −55‰ are indicative for aerobic methanotrophic bacteria (Bowman et al. 1993; Elvert and Niemann 2008; Hinrichs et al. 2003; Summons et al. 1994), and archaea-derived ether lipids such as archaeol and hydroxyarchaeol with very low $\delta^{13}C$-values between −104‰ and −94‰ are diagnostic for methanotrophic archaea (Blumenberg et al. 2004; Hinrichs et al. 1999). The lipid composition observed in the gut thus gives indirect insight into the community members present in the microbial mats thriving on the asphalts, indicating that both aerobic and anaerobic organisms are utilizing the methane released from the asphalts. Other more generic fatty acids are derived from bacteria that may be involved in sulfide oxidation. However, unfortunately no diagnostic lipid biomarker for these groups of organisms exists. The lipid biomass derived from the holothurian has relatively low carbon isotopic compositions, (e.g. cholesterol with −38‰), showing that these grazers mainly thrive on chemosynthetic biomass provided by the asphalt seep.

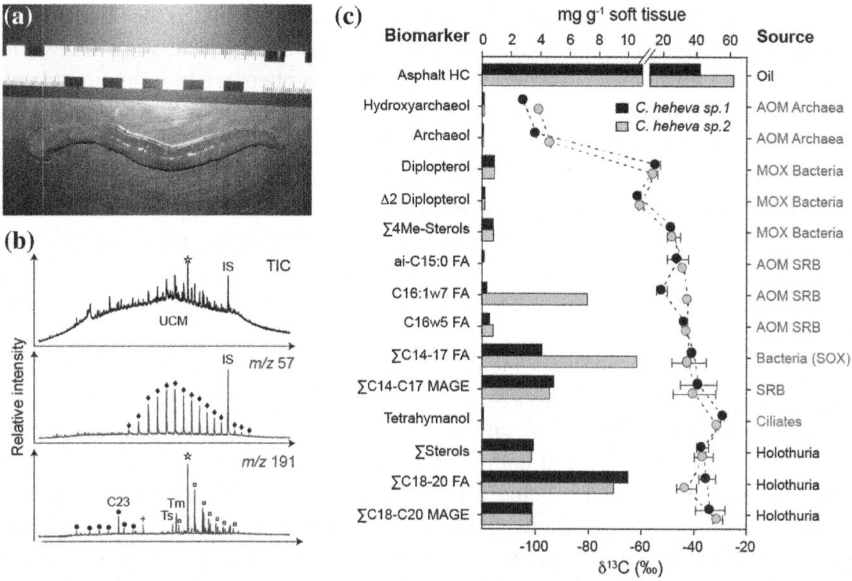

Fig. 6 Geochemical analysis of the gut content of asphalt-associated holothurian *Chiridota heheva*. **a** Image of one of the collected specimens of *Chiridota heheva*. **b** Total ion chromatogram (TIC), and extracted mass chromatograms *m/z* 57 and *m/z* 191 of the hydrocarbon fraction of one of the guts. Key of hydrocarbon compounds: UCM, unresolved complex mixture; IS, internal standard; filled diamonds, *n*-alkanes (C_{24}–C_{39}); filled circles, tricyclic terpanes (C_{19}–C_{25}); crosses, C_{24} tetracyclic terpane; Ts, trisnorneohopane; Tm, trisnorhopane; open squares, 17α21ß R and S hopane series (C_{28}–C_{35}); stars, norhopane. **c** Concentrations (represented by bars) and stable carbon isotopic compositions (represented by circles) of lipid biomarkers found in both holuthurian digestive tracks. Key of abbreviations: HC, hydrocarbons; FA, fatty acids; MAGE, monoalkyglycerolether; AOM, anaerobic oxidation of methane; MOX, aerobic methane oxidation; SRB, sulfate-reducing bacteria, SOX, sulfide-oxidizing bacteria. Organic compounds were extracted as described above and fractions were derivatized according to Hinrichs et al. (2000). Compounds were identified and quantified by gas chromatography coupled to mass spectrometry and flame ionization detection, respectively. Double bonds were determined according to Elvert et al. (2003). Compound-specific carbon isotopic compositions were measured by gas chromatography coupled to isotope ratio mass spectrometry as described by Schubotz et al. (2011b)

Summary and Outlook

The hydrocarbon province in the Southern Gulf of Mexico is characterized by seepage of heavy oil and large-scale deposition of asphalt beds on the seafloor. The asphalts themselves and co-emitted hydrocarbon gases provide an energy source for microorganisms, which form the basis of a highly diverse and exotic ecosystem at the Chapopote Knoll within the Bay of Campeche. In the sediment, and to some extent also within the asphalts, microbial degradation of hydrocarbons is coupled to sulfate reduction. Increased metabolic activity supported by the biodegradation by-product sulfide is evidenced by microbial mats and associated elevated oxygen consumption rates on top of the asphalts, as well as by symbiont-hosting tubeworms and mussels. Grazers such as decapods and sea cucumbers were found abundantly on top of the asphalts and sediments and often close to the occurrence of microbial mats. As hard settlement substrate, the asphalts provide a basis for the colonization by anemones, corals, bivalves, and mussels. These results demonstrate that the asphalt ecosystem has uniquely adapted to the occurrence of heavy oil, benefiting from the hydrocarbons as primary energy source. Lipids of typical seep microorganisms, including aerobic and anaerobic methanotrophs found in the gut of grazing cucumbers, exemplify the tight dependence of these organisms to seep activity.

In-depth molecular analyses including metagenomic investigations are required to grasp the diversity of organisms benefitting from the supply of the highly complex substrate mixture. This unique ecosystem harbors a large potential for the discovery of yet undiscovered pathways for the degradation of light and heavy oil components. Unfortunately, being a prolific and largely underexplored petroleum system makes the region around the Campeche Knolls an attractive target for oil and gas exploration, and it will be a challenge to preserve and protect this pristine and remote ecosystem for future exploration and scientific study.

Acknowledgements We thank the scientific crew and ROV team of R/V METEOR Expedition M67/2b. The Deutsche Forschungsgemeinschaft (DFG) and the Research Center/Excellence Cluster "The Ocean in the Earth System" are acknowledged for funding. Elva Escobar-Briones and Adriana Gaytán-Caballero are thanked for dissecting the holothurian and providing gut contents. Christiane Berndmeyer and Viola Beier are acknowledged for help in the laboratory and processing data and Kai-Uwe Hinrichs and Antje Boetius for providing laboratory facilities.

References

Aeckersberg F, Bak F, Widdel F (1991) Anaerobic oxidation of saturated hydrocarbons to CO_2 by a new type of sulfate-reducing bacterium. Arch Microbiol 156:5–14

Arellano SM, Lee OO, Lafi FF, Yang J, Wang Y, Young CM, Qian P-Y (2013) Deep sequencing of *Myxilla (Ectyomyxilla) methanophila*, an epibiotic sponge on cold-seep tubeworms, reveals methylotrophic, thiotrophic, and putative hydrocarbon-degrading microbial associations. Microb Ecol 65:450–461

Berdugo-Clavijo C, Gieg LM (2014) Conversion of crude oil to methane by a microbial consortium enriched from oil reservoir production waters. Front Microbiol 5:197

Bernard BB, Brooks JM, Sackett WM (1976) Natural gas seepage in the Gulf of Mexico. Earth Plant Sci Lett 31:48–54

Blumenberg M, Krüger M, Nauhaus K, Talbot HM, Oppermann BI, Seifert R, Pape T, Michaelis W (2004) Biosynthesis of hopanoids by sulfate-reducing bacteria (genus *Desulfovibrio*). Environ Microbiol 8:1220–1227

Boetius A, Wenzhöfer F (2013) Seafloor oxygen consumption fuelled by methane from cold seeps. Nat Geosci 6:725–734

Boetius A, Ravenschlag K, Schubert CJ, Rickert D, Widdel F, Gieseke A, Amann R, Jørgensen BB, Witte U, Pfannkuche O (2000) A marine microbial consortium apparently mediating anaerobic oxidation of methane. Nature 407:623–626

Borrel G, Adam PS, McKay LJ, Chen L-X, Sierra-García IN, Sieber CMK, Letourneur Q, Ghozlane A, Andersen GL, Li W-J, Hallam SJ, Muyzer G, de Oliveira VM, Inskeep WP, Banfield JF, Gribaldo S (2019) Wide diversity of methane and short-chain alkane metabolisms in uncultured archaea. Nat Microbiol 4:603–613

Bowman JP (2006) The methanotrophs–the families Methylococcaceae and Methylocystaceae. Prokaryotes 5:266–289

Bowman JP, Sly LI, Nichols PD, Hayward AC (1993) Revised taxonomy of the methanotrophs: Description of *Methylobacter* gen. nov., emendation of *Methylococcus*, validation of *Methylosinus* and *Methylocystis* species, and a proposal that the family Methylococcaceae includes only the Group I methanotrophs. Int J Syst Evol Microbiol 43:735–753

Brüning M, Sahling H, MacDonald IR, Ding F, Bohrmann G (2010) Origin, distribution, and alteration of asphalts at Chapopote Knoll, Southern Gulf of Mexico. Mar Petrol Geol 27:1093–1106

Bryant WR, Lugo J, Cordova C, Salvador A (1991) Physiography and bathymetry v. J. Boulder, Geological Society of America, Decade of North American Geology, The Gulf of Mexico Basin, pp 13–30

Coelho RR, Hovell I, de Mello Monte MB, Middea A, de Souza AL (2006) Characterisation of aliphatic chains in vacuum residues (VRs) of asphaltenes and resins using molecular modelling and FTIR techniques. Fuel Process Technol 87:325–333

Ding F, Spiess V, Brüning M, Fekete N, Keil H, Bohrmann G (2008) A conceptual model for hydrocarbon accumulation and seepage processes around Chapopote asphalt site, southern Gulf of Mexico: from high resolution seismic point of view. J Geophys Res Solid Earth 113: B08404

Dojka MA, Hugenholtz P, Haack SK, Pace NR (1998) Microbial diversity in a hydrocarbon- and chlorinated-solvent-contaminated aquifer undergoing intrinsic bioremediation. Appl Environ Microbiol 64:3869–3877

Dubilier N, Bergin C, Lott C (2008) Symbiotic diversity in marine animals: the art of harnessing chemosynthesis. Nat Rev Microbiol 6:725–740

Duperron S, Sibuet M, MacGregor BJ, Kuypers MM, Fisher CR, Dubilier N (2007) Diversity, relative abundance and metabolic potential of bacterial endosymbionts in three *Bathymodiolus* mussel species from cold seeps in the Gulf of Mexico. Environ Microbiol 9:1423–1438

Elvert M, Niemann H (2008) Occurrence of unusual steroids and hopanoids derived from aerobic methanotrophs at an active marine mud volcano. Org Geochem 39:167–177

Elvert M, Boetius A, Knittel K, Jørgensen BB (2003) Characterization of specific membrane fatty acids as chemotaxonomic markers for sulfate-reducing bacteria involved in anaerobic oxidation of methane. Geomicrobiol J 20:403–419

Feng G, Sun W, Zhang F, Karthik L, Li Z (2016) Inhabitancy of active Nitrosopumilus-like ammonia-oxidizing archaea and Nitrospira nitrite-oxidizing bacteria in the sponge *Theonella swinhoei*. Sci Rep 6:24966

Folmer O, Black M, Hoeh W, Lutz R, Vrijenhoek R (1994) DNA primers for amplification of mitochondrial cytochrome c oxidase subunit I from diverse metazoan invertebrates. Mol Mar Biol Biotechnol 3:294–299

Gutierrez T, Singleton DR, Berry D, Yang T, Aitken MD, Teske A (2013) Hydrocarbon-degrading bacteria enriched by the Deepwater Horizon oil spill identified by cultivation and DNA-SIP. ISME J 7:2091–2104

Guzman-Vega M, Mello M (1999) Origin of oil in the Sureste Basin, Mexico. AAPG Bull 83:1068–1095

Habe H, Omori T (2003) Genetics of polycyclic aromatic hydrocarbon metabolism in diverse aerobic bacteria. Biosci Biotechnol Biochem 67:225–243

Hallam SJ, Konstantinidis KT, Putnam N, Schleper C, Watanabe Y-I, Sugahara J, Preston C, de la Torre J, Richardson PM, DeLong EF (2006) Genomic analysis of the uncultivated marine crenarchaeote Cenarchaeum symbiosum. Proc Natl Acad Sci 103:18296–18301

Hanson RS, Hanson TE (1996) Methanotrophic bacteria. Microbiol Rev 60:439–471

Head IM, Jones DM, Röling WF (2006) Marine microorganisms make a meal of oil. Nat Rev Microbiol 4:173–182

Hinrichs K-U, Hayes JM, Sylva SP, Brewer PG, DeLong EF (1999) Methane-consuming archaebacteria in marine sediments. Nature 398:802–805

Hinrichs K-U, Summons RE, Orphan V, Sylva SP, Hayes JM (2000) Molecular and isotopic analysis of anaerobic methane-oxidizing communities in marine sediments. Org Geochem 31:1685–1701

Hinrichs K-U, Hmelo LR, Sylva SP (2003) Molecular fossil record of elevated methane levels in late Pleistocene coastal waters. Science 299:1214–1217

Hoffmann F, Radax R, Woebken D, Holtappels M, Lavik G, Rapp HT, Schläppy M-L, Schleper C, Kuypers MMM (2009) Complex nitrogen cycling in the sponge *Geodia barretti*. Environ Microbiol 11:2228–2243

Hostettler FD, Kvenvolden KA (1994) Geochemical changes in crude oil spilled from the Exxon Valdez supertanker into Prince William Sound, Alaska. Org Geochem 21:927–936

Jørgensen BB (1977) The sulfur cycle of a coastal marine sediment (Limfjorden, Denmark). Limnol Oceanogr 22:814–832

Kallmeyer J, Ferdelman TG, Weber A, Fossing H, Jørgensen BB (2004) A cold chromium distillation procedure for radiolabeled sulfide applied to sulfate reduction measurements. Limnol Oceanogr-Meth 2:171–180

Kane MD, Poulsen LK, Stahl DA (1993) Monitoring the enrichment and isolation of sulfate-reducing bacteria by using oligonucleotide hybridization probes designed from environmentally derived 16S rRNA sequences. Appl Environ Microbiol 59:682–686

Kasai Y, Kishira H, Harayama S (2002) Bacteria belonging to the genus *Cycloclasticus* play a primary role in the degradation of aromatic hydrocarbons released in a marine environment. Appl Environ Microbiol 68:5625–5633

Kerger BD, Nichols PD, Antworth CP, Sand W, Bock E, Cox JC, Langworthy TA, White DC (1986) Signature fatty acids in the polar lipids of acid-producing *Thiobacillus* spp.: methoxy, cyclopropyl, alpha-hydroxy-cyclopropyl and branched and normal monoenoic fatty acids. FEMS Microbiol Lett 38:67–77

Kleindienst S, Ramette A, Amann R, Knittel K (2012) Distribution and in situ abundance of sulfate-reducing bacteria in diverse marine hydrocarbon seep sediments. Environ Microbiol 14:2689–2710

Kleindienst S, Herbst FA, Stagars M, von Netzer F, von Bergen M, Seifert J, Peplies J, Amann R, Musat F, Lueders T, Knittel K (2014) Diverse sulfate-reducing bacteria of the *Desulfosarcina/Desulfococcus* clade are the key alkane degraders at marine seeps. ISME J 8:2029–2044

Knittel K, Boetius A, Lemke A, Eilers H, Lochte K, Pfannkuche O, Linke P, Amann R (2003) Activity, distribution, and diversity of sulfate reducers and other bacteria in sediments above gas hydrate (Cascadia margin, Oregon). Geomicrobiol J 20:269–294

Könneke M, Bernhard A, De la Torre J, Walker C, Waterbury J, Stahl D (2005) Isolation of an autotrophic ammonia-oxidizing marine archaeon. Nature 437:543–546

Kvenvolden KA, Cooper CK (2003) Natural seepage of crude oil into the marine environment. Geo Mar Lett 23:140–146

Larkin J, Henk MC, Aharon P (1994) Beggiatoa in microbial mats at hydrocarbon vents in the Gulf of Mexico and Warm Mineral Springs, Florida. Geo Mar Lett 14:97–103

Laso-Pérez R, Wegener G, Knittel K, Widdel F, Harding KJ, Krukenberg V, Meier DV, Richter M, Tegetmeyer HE, Riedel D, Richnow HH, Adrian L, Reemtsma T, Lechtenfeld OJ, Musat F (2016) Thermophilic archaea activate butane via alkyl-coenzyme M formation. Nature 539:396–401

Laso-Pérez R, Hahn C, van Vliet DM, Tegetmeyer HE, Schubotz F, Smit NT, Pape T, Sahling H, Bohrmann G, Boetius A (2019) Anaerobic degradation of non-methane alkanes by "Candidatus Methanoliparia" in hydrocarbon seeps of the Gulf of Mexico. MBio 10:e01814–01819

Lin Q, Mendelssohn IA (2012) Impacts and recovery of the Deepwater Horizon oil spill on vegetation structure and function of coastal salt marshes in the northern Gulf of Mexico. Environ Sci Technol 46:3737–3743

Lösekann T, Knittel K, Nadalig T, Fuchs B, Niemann H, Boetius A, Amann R (2007) Diversity and abundance of aerobic and anaerobic methane oxidizers at the Haakon Mosby Mud Volcano, Barents Sea. Appl Environ Microbiol 73:3348–3362

Ludwig W, Strunk O, Westram R, Richter L, Meier H, Yadhukumar N, Buchner A, Lai T, Steppi S, Jobb G, Forster W, Brettske I, Gerber S, Ginhart AW, Gross O, Grumann S, Hermann S, Jost R, Konig A, Liss T, Lussmann R, May M, Nonhoff B, Reichel B, Strehlow R, Stamatakis A, Stuckmann N, Vilbig A, Lenke M, Ludwig T, Bode A, Schleifer KH (2004) ARB: a software environment for sequence data. Nucleic Acids Res 32:1363–1371

Luter HM, Whalan S, Webster NS (2010) Exploring the role of microorganisms in the disease-like syndrome affecting the sponge *Ianthella basta*. Appl Environ Microbiol 76:5736–5744

MacDonald IR, Reilly Jr JE, Best SE, Venkataramaiah R, Sassen R, Guinasso Jr NL, Amos J (1996) Remote sensing inventory of active oil seeps and chemosynthetic communities in the northern Gulf of Mexico. In: Hydrocarbon Migration and Its Near-surface Expression: Outgrowth of the Aapg Hedberg Research Conference, Vancouver, British Columbia, April 24–28, 1994. American Association of Petroleum Geologists, p 27

MacDonald IR, Leifer I, Sassen R, Stine P, Mitchell R, Guinasso N (2002) Transfer of hydrocarbons from natural seeps to the water column and atmosphere. Geofluids 2:95–107

MacDonald IR, Bohrmann G, Escobar E, Abegg F, Blanchon P, Blinova V, Bruckmann W, Drews M, Eisenhauer A, Han X, Heeschen K, Meier F, Mortera C, Naehr T, Orcutt B, Bernard B, Brooks J, de Farago M (2004) Asphalt volcanism and chemosynthetic life in the Campeche Knolls, Gulf of Mexico. Science 304:999–1002

Magoon LB, Hudson TL, Cook HE (2001) AAPG Memoir 75, Chapter 4: Pimienta-tamabra—a giant supercharged petroleum system in the Southern Gulf of Mexico. Onshore and Offshore Mexico, pp 83–125

Marcon Y, Sahling H, MacDonald IR, Wintersteller P, dos Santos Ferreira C, Bohrmann G (2018) Slow volcanoes: the intriguing similarities between marine asphalt and basalt lavas. Oceanography 31(2):194–205

Massana R, Murray AE, Preston CM, DeLong EF (1997) Vertical distribution and phylogenetic characterization of marine planktonic Archaea in the Santa Barbara Channel. Appl Environ Microbiol 63:50–56

Musat F, Galushko A, Jacob J, Widdel F, Kube M, Reinhardt R, Wilkes H, Schink B, Rabus R (2009) Anaerobic degradation of naphthalene and 2-methylnaphthalene by strains of marine sulfate-reducing bacteria. Environ Microbiol 11:209–219

Muyzer G, Teske A, Wirsen CO, Jannasch HW (1995) Phylogenetic relationships of *Thiomicrospira* species and their identification in deep-sea hydrothermal vent samples by denaturing gradient gel electrophoresis of 16S rDNA fragments. Arch Microbiol 164:165–172

Nelson K, Fisher C (2000) Absence of cospeciation in deep-sea vestimentiferan tube worms and their bacterial endosymbionts. Symbiosis 28:1–15

Olajire A, Essien J (2014) Aerobic degradation of petroleum components by microbial consortia. J Petrol Environ Biotechnol 5:195

Orcutt BN, Joye SB, Kleindienst S, Knittel K, Ramette A, Reitz A, Samarkin V, Treude T, Boetius A (2010) Impact of natural oil and higher hydrocarbons on microbial diversity, distribution, and activity in Gulf of Mexico cold-seep sediments. Deep Sea Res II 57:2008–2021

Orphan VJ, Hinrichs KU, Ussler W, Paull CK, Taylor LT, Sylva SP, Hayes JM, Delong EF (2001) Comparative analysis of methane-oxidizing archaea and sulfate-reducing bacteria in anoxic marine sediments. Appl Environ Microbiol 67:1922–1934

Orphan VJ, House CH, Hinrichs KU, McKeegan KD, DeLong EF (2002) Multiple archaeal groups mediate methane oxidation in anoxic cold seep sediments. Proc Natl Acad Sci USA 99:7663–7668

Pape T, Hoffmann F, Quéric N-V, von Juterzenka K, Reitner J, Michaelis W (2006) Dense populations of Archaea associated with the demosponge *Tentorium semisuberites* Schmidt, 1870 from Arctic deep-waters. Polar Biol 29:662–667

Peters KE, Walters CC, Moldowan JM (2005) The biomarker guide. Cambridge University Press

Raggi L, Schubotz F, Hinrichs KU, Dubilier N, Petersen JM (2013) Bacterial symbionts of Bathymodiolus mussels and Escarpia tubeworms from Chapopote, an asphalt seep in the southern Gulf of Mexico. Environ Microbiol 15:1969–1987

Rahn O (1906) Ein Paraffin zersetzender Schimmelpilz. Zentralbl Bakteriol Parasitenk Infekt II. Abt 16:382–384

Rivers AR, Sharma S, Tringe SG, Martin J, Joye SB, Moran MA (2013) Transcriptional response of bathypelagic marine bacterioplankton to the Deepwater Horizon oil spill. ISME J 7:2315–2329

Rubin-Blum M, Antony CP, Borowski C, Sayavedra L, Pape T, Sahling H, Bohrmann G, Kleiner M, Redmond MC, Valentine DL (2017) Short-chain alkanes fuel mussel and sponge *Cycloclasticus* symbionts from deep-sea gas and oil seeps. Nat Microbiol 2:17093

Rütters H, Sass H, Cypionka H, Rullkötter J (2001) Monoalkylether phospholipids in the sulfate-reducing bacteria *Desulfosarcina variabilis* and *Desulforhabdus amnigenus*. Arch Microbiol 176:435–442

Sahling H, Borowski C, Escobar-Briones E, Gaytán-Caballero A, Hsu CW, Loher M, MacDonald IR, Marcon Y, Pape T, Römer M, Rubin-Blum M, Schubotz F, Smrzka D, Wegener G, Bohrmann G (2016) Massive asphalt deposits, oil seepage, and gas venting support abundant chemosynthetic communities at the Campeche Knolls, southern Gulf of Mexico. Biogeosciences 13:4491–4512

Salvador A (1991) Origin and development of the Gulf of Mexico Basin. In: Salvador A (ed) the Gulf of Mexico Basin, vol 2. Geology of North America, Decade of North American Geology, pp 389–444

Schubotz F (2009) Microbical community characterization and carbon turnover in methane-rich environments—case studies in the Gulf of Mexico and Black Sea. Deparment of Geosciences. University Bremen, Bremen, Bremen, p 222

Schubotz F, Lipp JS, Elvert M, Kasten S, Mollar XP, Zabel M, Bohrmann G, Hinrichs KU (2011a) Petroleum degradation and associated microbial signatures at the Chapopote asphalt volcano, Southern Gulf of Mexico. Geochim Cosmochim Acta 75:4377–4398

Schubotz F, Lipp JS, Elvert M, Hinrichs KU (2011b) Stable carbon isotopic compositions of intact polar lipids reveal complex carbon flow patterns among hydrocarbon degrading microbial communities at the Chapopote asphalt volcano. Geochim Cosmochim Acta 75:4399–4415

Söhngen N (1906) Über Bakterien, welche Methan als Kohlenstoffnahrung und Energiequelle gebrauchen. Zentrabl Bakteriol Parasitenk Infektionskr 15:513–517

Sommer S, Linke P, Pfannkuche O, Niemann H, Treude T (2010) Benthic respiration in a seep habitat dominated by dense beds of ampharetid polychaetes at the Hikurangi Margin (New Zealand). Mar Geol 272:223–232

Speight JG (1999) The chemical and physical structure of petroleum: effects on recovery operations. J Petrol Sci Eng 22:3–15

Summons RE, Jahnke LL, Roksandic Z (1994) Carbon isotopic fractionation in lipids from methanotrophic bacteria: relevance for interpretation of the geochemical record of biomarkers. Geochim Cosmochim Acta 58:2853–2863

Teske A, Ramsing NB, Habicht K, Fukui M, Küver J, Jørgensen BB, Cohen Y (1998) Sulfate-reducing bacteria and their activities in cyanobacterial mats of Solar Lake (Sinai, Egypt). Appl Environ Microbiol 64:2943–2951

Teske A, Hinrichs KU, Edgcomb V, de Vera Gomez A, Kysela D, Sylva SP, Sogin ML, Jannasch HW (2002) Microbial diversity of hydrothermal sediments in the Guaymas Basin: evidence for anaerobic methanotrophic communities. Appl Environ Microbiol 68:1994–2007

Treude T, Boetius A, Knittel K, Wallmann K, Jørgensen BB (2003) Anaerobic oxidation of methane above gas hydrates at Hydrate Ridge, NE Pacific Ocean. Mar Ecol Prog Ser 264:1–14

Van Beilen J, Li Z, Duetz W, Smits T, Witholt B (2003) Diversity of alkane hydroxylase systems in the environment. Oil Gas Sci Technol 58:427–440

Van Dover CL, Fry B (1994) Microorganisms as food resources at deep-sea hydrothermal vents. Limnol Oceanogr 39:51–57

Watanabe K (2001) Microorganisms relevant to bioremediation. Curr Opin Biotech 12:237–241

Watkins JS, Ladd JW, Buffler RT, Shaub FJ, Houston M, Worzel JL (1978) Occurrence and evolution of salt in deep Gulf of Mexico. Framework, facies, and oil-trapping characteristics of the upper continental margin. AAPG Studies in Geology 7:43–65

Wegener G, Shovitri M, Knittel K, Niemann H, Hovland M, Boetius A (2008) Biogeochemical processes and microbial diversity of the Gullfaks and Tommeliten methane seeps (Northern North Sea). Biogeosciences 5:1127–1144

Wenzhöfer F, Glud RN (2002) Benthic carbon mineralization in the Atlantic: a synthesis based on in situ data from the last decade. Deep-Sea Res I 49:1255–1279

Widdel F, Rabus R (2001) Anaerobic biodegradation of saturated and aromatic hydrocarbons. Curr Opin Biotech 12:259–276

Zengler K, Richnow HH, Rossello-Mora R, Michaelis W, Widdel F (1999) Methane formation from long-chain alkanes by anaerobic microorganisms. Nature 401:266–269

Zhou J, Bruns MA, Tiedje JM (1996) DNA recovery from soils of diverse composition. Appl Environ Microbiol 62:316–322

Chapter 6
Archaea in Mediterranean Sea Cold Seep Sediments and Brine Pools

Cassandre S. Lazar

Abstract Cold seeps host intense and complex biochemical processes, in particular methane and sulfur cycling. Microorganisms are key players in these habitats, producing or oxidizing methane, and reducing sulfate. Mediterranean cold seep mud volcanoes are natural laboratories allowing to study how reducing fluids from different volcanoes with distinct connections to the marine subsurface, influence composition and activities of benthic microbial communities. Methane-producing and-oxidizing archaea were detected in mud volcanoes associated with different seepage intensities, in hypersaline brine-impacted sediments, as well as in pockmark structures. Because the muds and fluids expelled from the center of the mud volcanoes ascend from deep sources, they could be windows into the deep biosphere, allowing a glimpse into the diversity of communities surviving in the deep subsurface.

Introduction

Reduced habitats in the sediments of oceans or seas are unique environments harboring a vast biodiversity. Hydrothermal sources were one of the first ecosystems to be discovered in deep-sea habitats (Corliss et al. 1979). A few years later, a new type of ecosystem was uncovered, in regions of the seafloor where life seemed unlikely (Paull et al. 1984; Suess et al. 1985): cold, dark, anoxic zones with fluids and muds ascending from deep sources were shown to sustain life. They were called cold seeps. These regions of the marine subsurface are referred to as cold, opposing the extremely hot fluids found in hydrothermal vents. Nonetheless, fluids reaching the seafloor in cold seeps are usually warmer than the bottom seawater. In

C. S. Lazar (✉)
University of Quebec in Montreal (UQAM), Montreal, Canada
e-mail: lazar.cassandre@uqam.ca

A. Teske and V. Carvalho (eds.), *Marine Hydrocarbon Seeps*,
Springer Oceanography, https://doi.org/10.1007/978-3-030-34827-4_6

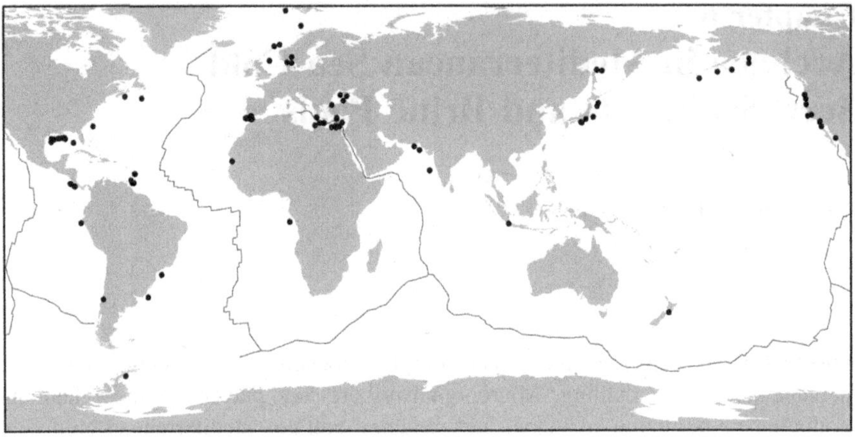

Fig. 1 Worldwide distribution of cold seeps (Lazar 2010)

these ecosystems, and in the absence of sunlight, chemosynthetic organisms rule, using inorganic compounds as energy and carbon sources. Cold seeps are distributed globally, situated mostly in passive oceanic margins (Fig. 1).

Geological Structures in Cold Seeps

Continental margins are frontiers between a continental and oceanic plate. Here, two types of margins can be found: active and passive. Many active margins are subduction zones at tectonic plate boundaries, accompanied by intense volcanic and seismic activities. Passive margins are stable zones located at inactive plate boundaries, at the transition between the terrestrial and oceanic crusts. Fluids circulate in the marine sedimentary layers of continental margins. The principal force behind fluid circulation is the compaction of sediments in the marine subsurface, the inherent dynamics of buried salt layers (see Chap. 4), and—at active margins—tectonic plate convergence movements, which lead to overpressure phenomena (Levin 2005). Thus, fluids and muds ascend along porous and permeable layers such as fractures, fissures, or mud layers. The over-pressurized fluids contain organic hydrocarbons such as methane or petroleum compounds.

When fluids and muds reach the seafloor and are expulsed, emissions from deep sources are often associated with spectacular geological formations, such as mud volcanoes, pockmarks, or submarine slides (see also Chap. 7). Mud volcanoes are generally circular with diameters ranging from meters to kilometers. They possess an active center, from which fluids and mud are emitted. The fluids are rich in dissolved gases, predominantly methane, but also carbon dioxide, sulfide, and ammonia. Mud volcanoes can show violent and intense eruptions, but mostly their expansion progresses by gradually emitting semi-liquid muds.

Pockmarks are marine conical depressions of variable diameter (between 1 and 100 m) and depth (between 1 and 10 m). Pockmarks are the visible result of degassing processes from deep sources that are impacting shallow sediment layers, on the seafloor. They are most often situated close to mud volcanos, along fault lines, and at fluid and mud emission sites. Gases ascending in mud volcanoes lead to the formation of specific channels through mud and pelagic sediments. Shallow reservoirs can then form when these pathways are sealed by impermeable mud, and filled up with gases ascending from deep sources. Pressure then increases in these pockets, until a triggering event (such as an earthquake) leads to the expulsion of sediments and gases, thus creating a pockmark. Once the pathway is created, it will become the preferential route for gases towards the seafloor. Additional conduits can be created if the principal channel is blocked. This can explain the emergence of groups of pockmarks within close proximity to each other. Pockmarks can remain active, slowly emitting gases during long periods, or they can be quiescent (Dimitrov and Woodside 2003).

Geochemical Conditions in Cold Seeps

Marine sediments produce 85–300 Tg of methane per year (Knittel and Boetius 2009) and store methane in marine sediments, as free or dissolved gas, or as a solid in the form of gas hydrates. In gas hydrates, methane molecules are surrounded by a crystalline cage of water molecules. They form naturally when the concentration of dissolved methane reaches saturation, and in specific pressure and temperature conditions (Brown et al. 2006). In marine sediments, methane can either be produced under biogenic or abiotic conditions. In the marine subsurface, however, methane is thought to be principally produced by microorganisms (Kvenvolden 1995).

Sulfide also plays a big role in marine sediment biogeochemical cycles (Omoregie et al. 2008). Along with methane, sulfide can be used as electron donor by a wide range of microorganisms. Hydrogen sulfide is produced during anaerobic degradation of organic matter, coupled with the reduction of sulfate (Sahling et al. 2002). Anaerobic remineralization of organic matter by sulfate reduction, and sulfide oxidation by chemosynthetic microorganisms, are two major biogeochemical processes linking the carbon and sulfur cycles in marine sediments (details for these microbial processes can also be found in Chaps. 1, 2, 5, and 9).

Marine organic matter is principally produced by living organisms such as phytoplankton in the water column, and ultimately settles on the seafloor as particulate organic carbon, forming a fine layer of detritus at the water-sediment interface on the seafloor. Transitioning from fresh organic matter into older, more complex structures during microbial degradation, it gets progressively buried in the marine subsurface. In sediments, microbes obtain energy by oxidation of organic matter using an external electron acceptor. From a thermodynamic point of view, oxygen is the most favorable acceptor in aerobic habitats. As layers of sediments are gradually reduced, a succession of microbial respiratory types is established. After oxygen has been depleted by aerobic organisms, anaerobic microorganisms use other electron acceptors to mineralize organic matter. Nitrate, manganese and iron oxides are the most energetically favorable. However, they are usually depleted within the first centimeters. Sulfate is present as one of the major ions in seawater (concentration ca. 28 mM), and permeates to the sedimentary subsurface by diffusion and bioturbation processes, where it is used as electron acceptor by sulfate-reducing bacteria (SRB) when nitrate, manganese and iron compounds are exhausted. In deeper anoxic layers, where even sulfate is depleted, organic carbon is no longer oxidized, and its degradation via fermentative pathways and methanogenic microorganisms results in the accumulation of methane (Jørgensen and Kasten 2006). As methane slowly diffuses towards the sediment layers where sulfate becomes available, it is microbially converted to carbon dioxide by specialized microbial communities performing anaerobic, sulfate-dependent methane oxidation (AOM). This crucial microbial process occurs globally in marine sediments (see also Chap. 1). The sediment layers where sulfate and methane overlap, and are concurrently consumed by anaerobic methane oxidation, are termed the sulfate-methane transition zone (SMTZ). Below the SMTZ, when sulfate is no longer available, methanogenesis rules. Here, methane production is the final step of organic matter degradation, and methane accumulates. For this reason, methane is the dominant fossil hydrocarbon compound of the planet (Knittel and Boetius 2009).

Mud Volcanoes in the Mediterranean Sea

Over 200 mud volcanoes have been found along the northern flank of the Mediterranean Ridge in the eastern Mediterranean Sea (Charlou et al. 2003). The formation of the Mediterranean Ridge is linked to the collision between the African and Eurasian tectonic plates, resulting in intense faulting (Haese et al. 2006). Cold seeps are also found further east along the Cyprus Arc within the Nile deep-sea fan (Fig. 2). Specifically, two areas that are rich in biological and geological diversity have been the focus of multidisciplinary projects and sampling cruises since the 1990s: the Olimpi area south of Crete where mud volcanoes are widespread, and the Anaximander Mountains south of Turkey where eruptions are associated with active tectonic deformation (Charlou et al. 2003). The Anaximander Mountains are detached continental blocks, and are not associated with the Mediterranean Ridge as the Olimpi area (Zitter et al. 2005). They are associated with very active tectonic deformation (Charlou et al. 2003), and show evidence of active fluid seepage (Haese et al. 2006).

Another difference between both areas is the presence of Messinian evaporites deposited in basins in the Olimpi area, which are absent in the Anaximander Mountains. Therefore, brine seepage is often associated with mud volcanoes in the Olimpi area, and in contrast is never observed in the Anaximander Mountains. Mud

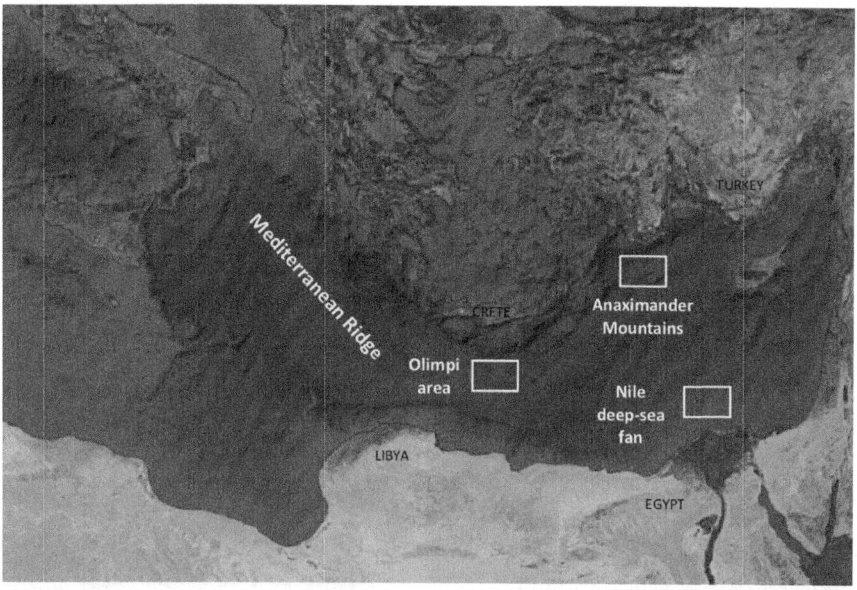

Fig. 2 Location of the Olimpi area, the Anaximander Mountains, and the Nile deep-sea fan in the Mediterranean Sea. Images: Medeco 2007, copyright Ifremer

volcanoes in the Olimpi area show a wide range in geological morphologies, diameters, salinity, and associated fauna. Here, several volcanoes named Napoli, Milano, Moscow, and Maidstone have been mapped and studied (Charlou et al. 2003). In the Anaximander Mountains, three major mud volcanoes—Amsterdam, Kazan, and Kula—have been identified and investigated.

In October 2007, the IFREMER sampling cruise Medeco, set out to study biochemical and biological features in Mediterranean cold seep sediments. Here, case studies of selected mud volcanoes, Napoli from the Olimpi area, and Amsterdam from the Anaximander Mountains, are presented. Given the importance of methanogenesis and anaerobic methane-oxidation, this survey focuses on the diversity of archaeal communities in these selected Mediterranean cold seep habitats. In situ images taken with the ROV *Victor* at both mud volcanoes are reproduced in this chapter, and unveil characteristic structures and fauna associated with mud volcanoes—such images are for the most part rarely disclosed in published literature, despite their highly valuable contribution to the understanding of life and geological features in Mediterranean cold seep environments.

The Amsterdam mud volcano is characterized by a rough seafloor topography with many depressions and scars (Fig. 3a), and the seafloor features abundant carbonate crusts interspersed with soft mud (Fig. 3b; Olu-Le-Roy et al. 2004). High methane concentrations were measured in the overlying water, while the methane originated from subsurface sediments and expelled mud in the central part of the mud volcano (Lykousis et al. 2009). Degassing at the Amsterdam mud volcano is linked with high turbidity, and hence the Amsterdam mud volcano is characterized as a location with high levels of methane venting and extensive turbid fluid expulsion. The occurrence and distribution of seep fauna suggests high chemosynthetic activity (Fig. 3c–f).

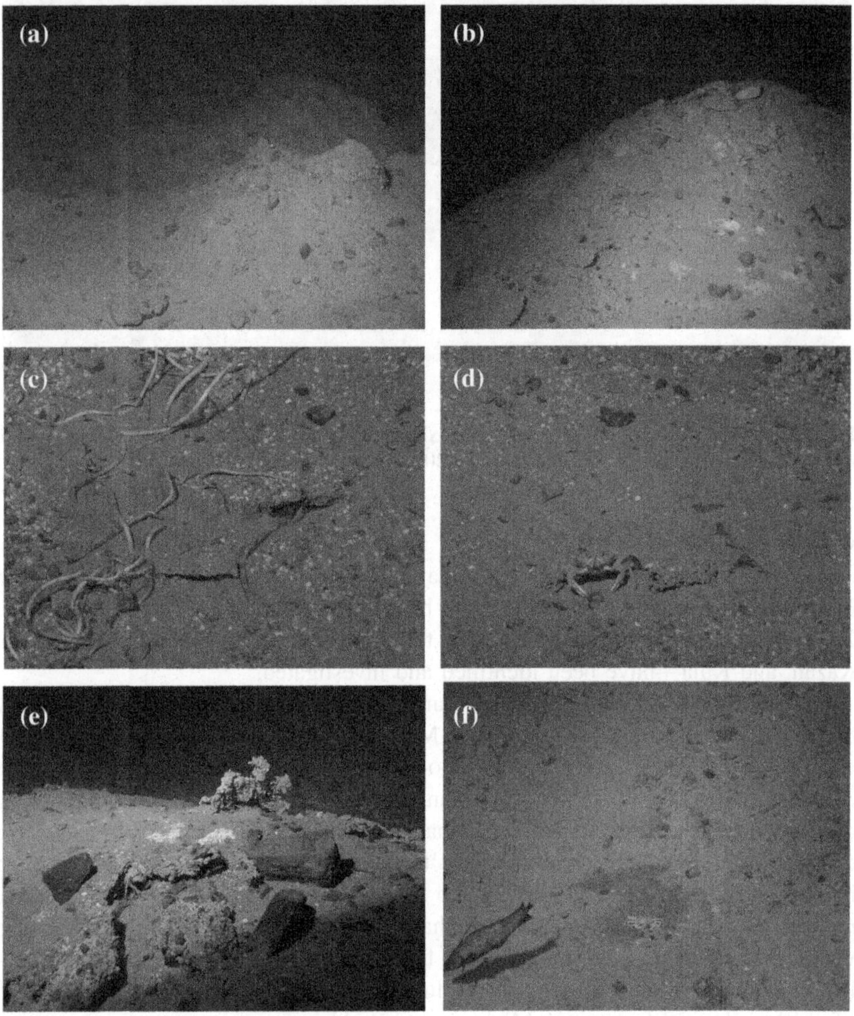

Fig. 3 In situ ROV photographs of the Amsterdam mud volcano, taken during the Medeco cruise in October 2007. **a** and **b** Show the characteristic "hump" morphology and carbonate crusts of sediment-covered methane hydrates; **c–f** show characteristic faunal assemblages with tubeworms (**c**), crustaceans (**d**, **e** and **f**), small microbial mats and possibly biogenic sediment accretions (**e**), as well as benthic fish (**f**) associated with the mud volcano. Images: Medeco 2007, copyright Ifremer

The Napoli mud volcano is a circular dome associated with ascending hypersaline fluids, and is characterized by brine pools and lakes with a diameter size ranging from centimeters to meters (Fig. 4a, b; Charlou et al. 2003). Fauna was observed on the summit (active inner zone, Fig. 4c–g), whereas the outer zone hosted mostly dead animals (Fig. 4h), suggesting that chemosynthetic activity is moderate due to brines and mild fluid flows (Olu-Le-Roy et al. 2004). Most of the gas venting from Napoli is composed of methane, mainly of biogenic origin.

Microorganisms in the Center of the Amsterdam Mud Volcano

Sediments at the center of the Amsterdam mud volcano were sampled during the Medeco cruise, using two gravity cores reaching 650 cm below the seafloor (cmbsf). These sediments were characterized by relatively high temperatures in deeper layers, i.e. 21–36 °C at 600 cmbsf compared to 14 °C at the seabed. Acetate concentrations were extremely high in the sediment pore waters at the center of the Amsterdam mud volcano. The SMTZ was assumed to be situated at around 50 cmbsf.

Molecular surveys (based on 16S rRNA gene sequencing), methanogenesis activity and geochemical measurements highlighted sediments layers within the center of the mud volcano, with distinct methanogenic archaeal communities and substrates (Lazar et al. 2012). In the sulfate-rich layers (0–50 cmbsf), members of the methylotrophic family *Methanosarcinaceae* as well as from the genus *Methanococcoides* were detected, which coincided with a peak in methylotrophic methanogenic activities. In layers between 50 and 100 cmbsf, characterized by high acetate concentrations, acetoclastic and methylotrophic methanogenic activities were measured, however, acetate-based methanogenesis was 10 times higher. In layers between 100 and 200 cmbsf, methylotrophic and hydrogenotrophic methanogenesis were measured, which could be carried out by detected members of the hydrogenotrophic genus *Methanogenium*, and by relatives of the methylotrophic genus *Methanosarcina*. Numerous sequences were also affiliated with ANME-2 lineages within the *Methanosarcinales*. From 200–350 cmbsf, acetoclastic methanogenesis was the dominant methane-producing pathway. Methanogenesis rates using methanol and H_2/CO_2 were also detected, but were ca. 10 times lower (Lazar et al. 2012). This shift from hydrogenotrophic to acetoclastic methanogenesis is probably caused by geothermal heating of mud volcano sediments. In laboratory experiments using different types of sediments, acetate could be generated from sedimentary organic matter at relatively moderate temperatures (between 10 and 60 °C); the same process can release acetate in deep subsurface sediments that are heated by a natural geothermal gradient (Wellsbury et al. 1997). Thermal gradient measurements at the Amsterdam mud volcano showed that in situ temperatures increased with depth, reaching 20–25 °C at 350 cmbsf. Thus, increasing temperatures result in increased pore water acetate concentrations, which is the substrate for acetoclastic

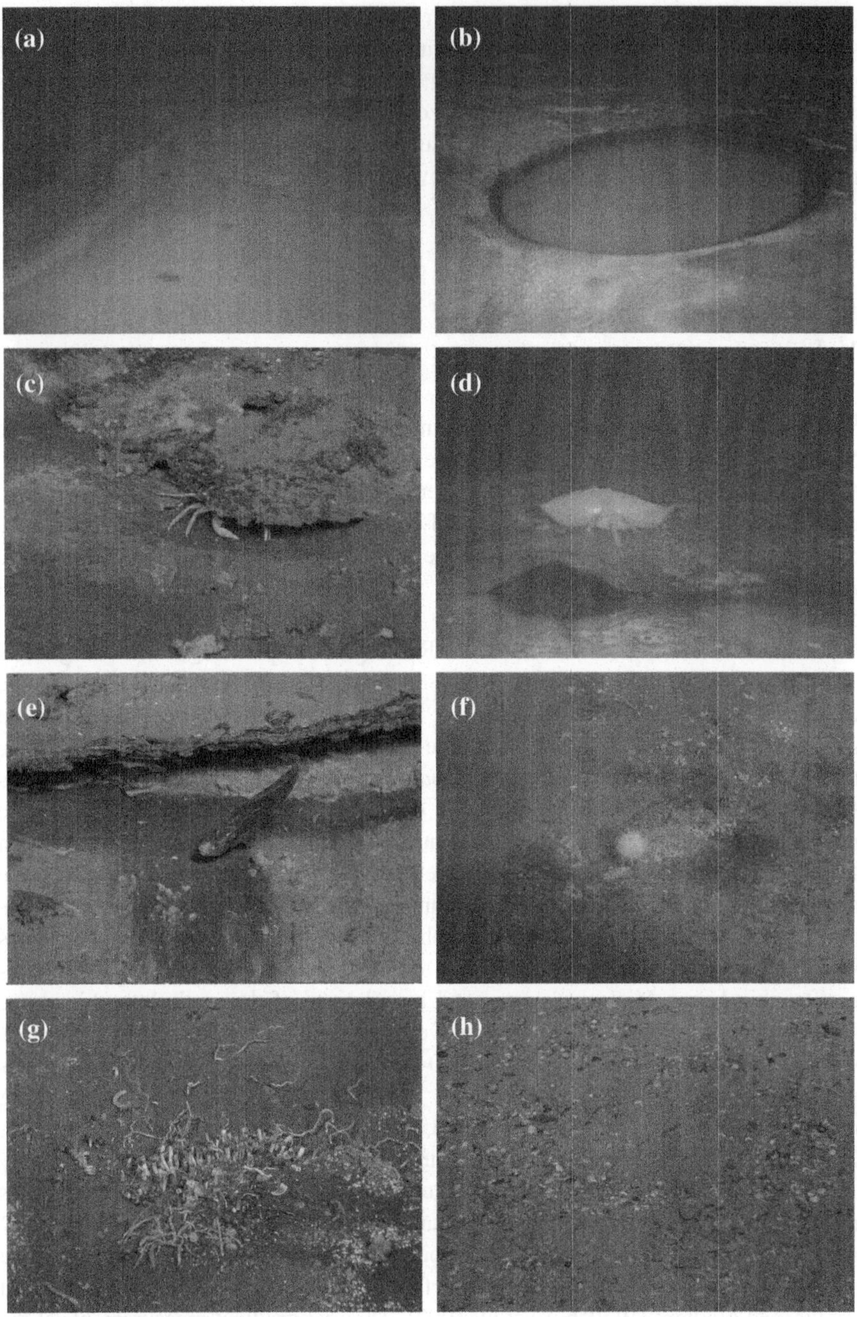
(a)
(b)
(c)
(d)
(e)
(f)
(g)
(h)

◂**Fig. 4 In situ ROV photographs of the Napoli mud volcano, taken during the Medeco cruise in October 2010.** **a** and **b** show brine flows and pools associated with Napoli; **c–g** show fauna associated with the mud volcano. **c** A crab hiding under a slab of carbonate crust, surrounded by dark-grey, sulfidic sediments; **d** A skate gliding over an extensive area of microbial mats; **e** The entire benthic food chain in a single photo: a benthic fish in front of a surface-breaching carbonate crust, over dark-grey sulfidic sediments that are flanked by feathery-white microbial mats (probably sulfur-oxidizing *Beggiatoaceae*, Chap. 8) and populated by infaunal worms; **f** A sea urchin grazing on a mosaic of sulfidic grey sediments, and orange- or rust-colored sediment stained by iron oxidation at the sediment surface; **g** A tubeworm colony, conspicuous red gills emerge like small plumes at the top of each habited chitin tube; **h** Clam shells and possibly worm tube fragments in the outer zone of the mud volcano. Images: Medeco 2007, copyright Ifremer

methanogenesis. Furthermore, in methane-saturated sediment, acetogenic bacteria could outcompete hydrogenotrophic methanogens for bicarbonates (Heuer et al. 2009). Below 350 cmbsf, acetoclastic methanogenesis ceased and methane was produced exclusively from methylamines. Sequences related to *Methanosarcinales* were detected in sediments from 550 to 600 cmbsf suggesting that these methanogens utilize methylamines for methane production (Lazar et al. 2012). Methylotrophic methanogenesis was measured throughout the sediment core, highlighting that these methanogens were able to find alternative methylated substrates to support their growth in different geochemical environments.

Members of the ANME-1 clade, which catalyze the anaerobic oxidation of methane (Knittel and Boetius 2009), were detected in the SMTZ, indicating that AOM could occur at 50 cmbsf in the center of the Amsterdam mud volcano. Furthermore, ANME-1 were present from the sediment-water interface down to 400 cmbsf, and ANME-2c occurred from 100 down to 600 cmbsf depth. In other studies, the ANME-2c subgroup was mostly detected in environments with low fluid fluxes (Mills et al. 2003, 2004; Inagaki et al. 2004; Fang et al. 2006), and high methane concentrations (Wegener et al. 2008). In high-flow regimes, ANME-1 could outgrow and outcompete ANME-2 for resources (Girguis et al. 2005). Some sequences obtained from sediments of the Amsterdam mud volcano belonged to the *mcrA* a/b group (ANME-1; Hallam et al. 2003), and formed a distinct cluster affiliated with clones retrieved from sediments of the Kazan mud volcano (with 92–98% of similarity), which is also part of the Anaximander Mountains. Environmental conditions encountered in sediments of the Amsterdam mud volcano and generally in the Anaximander Mountains (i.e., high methane fluxes, high temperatures in deep sediments, presence of methane hydrates), probably determined specific ecological niches for adapted microbial populations.

In summary, archaeal methanogenic and methanotrophic populations co-occurred in sediments at the center of the Amsterdam mud volcano, and acetoclastic methanogenesis was the dominant methanogenic pathway identified here. High acetate concentrations could be explained by abiotic production in the deep hot reservoir of the mud volcano. High ascending methane flows throughout sediments at the center of the Amsterdam mud volcano could prevent downwards diffusion of sulfate from the water column, explaining the lower sulfate concentrations measured below 50 cmbsf.

Microorganisms in the Center of the Napoli Mud Volcano

Sediments were sampled during the Medeco cruise at the Napoli mud volcano, using two gravity cores reaching 120–160 cmbsf. Temperature measurements in the sediment cores recovered from near the center of the mud volcano indicated an average of 14 °C, with a gradient of 100 °C/km. This value remains low compared to the Amsterdam mud volcano, supporting the evidence that Napoli is a cold mud volcano with moderate fluid flow rates and activities. Chloride porewater concentrations reached amounts 10 times higher than seawater concentrations, confirming that ascending fluids are brine-enriched. Sulfate was present both in the shallow layers and in the deeper depths, indicating that the center of Napoli is influenced by a deep brine incursion as well. Below 22 cmbsf, sulfate concentrations never reached depletion, suggesting that sulfate is also diffusing upwards from the deep brine sources. Methane was increasing below 60 cm, and reached maximum concentrations at 130 cmbsf. Because of the irregular aspect of the sulfate concentrations, it was impossible to situate the SMTZ, although it was proposed to be located between 20 and 40 cmbsf, where methane concentrations were minimal (Lazar et al. 2011a).

Methanogenic activities in the center of Napoli were in general lower than rates measured in Mediterranean brines and in the brine-impacted Gulf of Mexico (Joye et al. 1999; Daffonchio et al. 2006), possibly indicating that ascending fluids could hinder efficient methanogenesis. Methylamine and methanol-based methanogenic activities were measured in the sulfate-rich sediments where sulfate-reduction activities are high. Generally, in these layers, sulfate-reducing bacteria outcompete methanogens for substrates such as H_2 and acetate (Holmer and Kristensen 1994). However, SRB do not compete for methylated substrates, which is why it is not unusual to find methylotrophic methanogens in sulfate-rich marine sediments (Roussel et al. 2009). Between 40 and 100 cmbsf, where sulfate and methane were detected, and acetate concentrations were highest, methanogenic activities based on H_2/CO_2, acetate, and methylotrophic compounds were detected. Hydrogenotrophic methanogenesis was the dominant methane formation pathway at this depth. Below 100 cmbsf, only acetoclastic and methylotrophic methanogenesis were detected, with a dominance of acetate-based methane production. Thus, in these layers, methanogens shifted from using H_2/CO_2 to acetate as a methanogenic substrate.

Two major factors are known to influence methanogenic activity: temperature and substrate availability. Temperature shifts from 4 to 20 °C can induce a change from acetate-based to H_2/CO_2-based methanogenesis (Glissmann et al. 2004). Hydrogenotrophic methanogenesis is limited by the supply of H_2, which is usually linked to H_2-producing syntrophs sensitive to low temperatures. Temperatures measured at 14 °C in the Napoli sediments could be one factor influencing hydrogenotrophic over acetoclastic methanogens. Also, traditionally, high salinities favor H_2/CO_2-using methanogens, because acetate-using methanogens cannot tolerate these conditions (Oren 1999). However, below 120 cmbsf acetate-based methanogenesis was detected whereas H_2/CO_2-based activities were not. Since acetoclastic methanogenesis was reported in a brine-impacted mud volcano in the Gulf of Mexico (Joye et al. 2009), it seems likely some acetoclastic methanogens have adapted to hypersaline environments.

A dominant portion of the *mcrA* sequences were affiliated with ANME archaea. Archaea belonging to the ANME-1 group were detected in all sediment layers, and archaea belonging to the ANME-2a group were detected below 40 cmbsf. As previously suggested, in high flow regimes such as the habitats found in the center of a mud volcano (Girguis et al. 2005), ANME-1 archaea could outgrow and outcompete ANME-2 archaea for resources. Members of the *mcrA* a/b group cluster (ANME-1; Hallam et al. 2003) were recovered from most sediment layers at the center of Napoli, and contained sequences closely affiliated with ANME-1 genes from hypersaline sediments of the Gulf of Mexico (Lloyd et al. 2006). These archaea were proposed to be a subpopulation adapted to high-salt environments, which could explain their dominance in the brine-impacted Napoli volcano. *McrA* sequences affiliated with the methanogenic genus *Methanococcoides* at depths between 40 and 60 cmbsf, whereas sequences affiliating with ANME-3 archaea were detected below between 60 and 80 cmbsf. As previously suggested in methane-rich sediments of Monterey Canyon in the Gulf of Mexico (Goffredi et al. 2008), this distribution could show that ANME-3 are adapted to niches with a low availability of carbon sources. Overall, detection of diverse ANME archaea based on the *mcrA* functional genes, strongly suggests that AOM is occurring in the hypersaline sediments of Napoli.

The brine-impacted sediments of the center of the Napoli mud volcano represent a dynamic ecological niche for methanogens, ANME, and other archaea that have adapted to variations in fluid flow and high salinity. The presence of active methanogens in the upper 100 cm of fluidized mud from the volcano could indicate transport through the ascending fluids from deeper reservoirs. Hence, methanogenesis in these sediment layers might be directly connected and affected by the deep biosphere.

Microorganisms in Microbial Mats Associated with the Napoli Mud Volcano

The Olimpi area is a dynamic environment, with active seepage of mud, fluids, and brines. During the Miocene era, the Mediterranean Sea was isolated from the Atlantic Ocean leading to massive evaporation and deposition of evaporites in deeper basins. During the Messinian salinity crisis, evaporites were deposited in Mediterranean basins (Zitter et al. 2005). Thus, continuous dissolution of the evaporitic layer and upward diffusion of brines are occurring in the Olimpi area (Dählmann and de Lange 2003). Subsurface brines intrude into the seafloor of the mud volcanoes within the Olimpi area, creating brine pools and lakes with diameters ranging from centimeters to meters. Dense microbial mats of filamentous

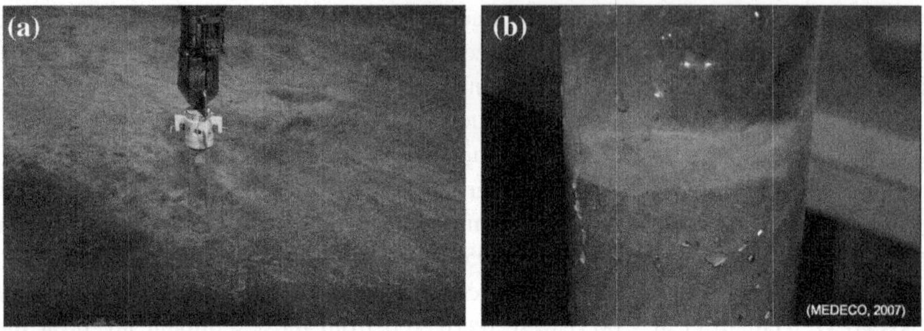

Fig. 5 **a** ROV image of the sampling of orange microbial mats associated with the Napoli mud volcano; **b** Freshly recovered sediment core, covered by a mat of orange microbial filaments. Images: Medeco 2007, copyright Ifremer

bacteria can be found on the seafloor above Mediterranean cold seep sediments, visible to the naked eye (Fig. 5a, b). These mats consist mainly of filamentous sulfur-oxidizing bacteria (Heijs et al. 2005; Grünke et al. 2011). Therefore, sediments underlying these mats provide alternative niches for diverse microbial communities adapted to dynamic changes of fluid flow regimes.

During the Medeco cruise, hypersaline sediments underlying an orange-pigmented microbial mat at the Napoli mud volcano were sampled (Lazar et al. 2011b). These short (10 cm) cores were subsampled every 2 cm. Chloride concentrations reached 1200 mM at 10 cmbsf, which is more than twice the seawater concentrations, reflecting the high salinity of the upflowing fluids. In the first 2 cm below the mat, sulfate concentrations were higher (30 mM) than in the overlying seawater, most likely reflecting active production of sulfate by the filamentous bacteria within the mat. The bacterial filaments penetrated the first few centimeters of the sediments, probably diffusing sulfate into the deeper layers. It is thus likely that the SMTZ was not located within the first 10 cm.

No *mcrA* sequences were detected in the first 2 cm below the mat. This could be explained by an oxygenation of the shallow sediment layers via the motility of the bacterial mat filaments. In layers between 2 and 4 cmbsf, ANME-2a archaea were dominant, whereas in the deeper 8–10 cmbsf layers, ANME-1 archaea were dominant. As discussed previously, ANME-1 could be able to outgrow and outcompete ANME-2 for resources (Girguis et al. 2005). Also, ANME-1 have been suggested to be more sensitive to oxygen than ANME-2 (Knittel et al. 2005), which could explain the ANME community patterns observed in the sediments underlying the microbial map of the Napoli volcano, i.e., the tolerance of ANME-2 for less reducing conditions, allowing them to survive near the seawater interface, whereas ANME-1 require fully reducing, sulfidic conditions found in deeper sediment layers (Rossel et al. 2011).

RNA-derived 16S rRNA sequences were analyzed from sediments depths 0–4 cmbsf. Two groups were found to be active and dominant in these layers: the MBG-D archaea, and the Rice Cluster V (RC-V) archaea, both belonging to uncultured lineages. Genomes affiliated with the RC-V have been reconstructed, predicting a heterotrophic metabolism based on the utilization of complex carbohydrates (Lazar et al. 2017). These shallow sediment layers had a high organic matter content, which could be derived from organic detritus resulting from microbial activities in the mat. Thus, both the MBG-D and RC-V archaea were most likely actively using extracellular detrital matter as source of carbon and energy in the shallow sediment layers.

Neither of the studies investigating the brine-impacted sediments within or associated with the Napoli mud volcano detected halophilic or halotolerant archaea, despite the extremely high chloride concentrations. The same observation was made in hypersaline sediments from the Gulf of Mexico (Joye et al. 2009). Possibly, some of the yet-unidentified archaea detected in the Napoli sediments represent as yet unknown halophilic or halotolerant organisms.

Microorganisms in Pockmark Sediments of the Nile Deep Sea Fan

The Nile Deep sea fan is characterized, by the presence of numerous fluid emission sites at the seafloor. Influenced by factors such as temperature, pressure, and important tectonic constraints, the thick layer of sediments constituting the Egyptian continental margin (more than 150 million years old) is constantly deformed and releases fluids and brines that are rich in gases and hydrocarbons. After the Messinian crisis, and successive deposits of thick sediments and evaporite layers, the tectonic activities increased in Egypt (Dupré et al. 2007). In the western and eastern hydrocarbon-rich regions of the Nile delta, fluid migrations through tectonic faults are very intense (Dupré et al. 2007).

Pockmarks on the seafloor are evidence for regions where cold seeps are emitted. They are very abundant in the central and western regions of the Nile delta. In the central regions, extended pockmark fields are found in a major sediment destabilization area. The pockmarks found in the different regions of the Nile deep sea fan are active sites associated with methane emissions and carbonate crusts, while the carbon is derived from the activity of anaerobic methane oxidizing microbes (Gontharet et al. 2007).

During the Medeco cruise, sediment cores were sampled in a pockmark area also covered with carbonate crusts, in the southern part of the Nile deep sea fan (Lazar 2010). Two push cores were recovered in two sites where the seafloor was covered with white granules (Fig. 6a, b).

The cores displayed intense degassing, as shown by bubbling in the sediment, indicating active fluid and gas release (Fig. 6c). Porewater sulfate concentrations were depleted below 8 cmbsf, which is either evidence for rapid sulfate consumption, or, an upward migration of sulfate-free fluids. Microbial mats or tubeworms could not be found, neither macroscopically nor during microscopic

Fig. 6 In situ ROV photos of extensive areas of white seafloor deposits in the Nile deep sea fan (**a**), and sampling of the pockmark-associated sediments in this area (**b**). Close-up of freshly recovered gas-rich sediment push core (**c**). Images: Medeco 2007, copyright Ifremer

observations of the surface sediment. Intense fluid and gas release at the seafloor in this area could have prevented the establishment of a seabed chemosynthetic community.

The core from site 2A contained 10 mM of sulfate in the top 2 cm, whereas the one from site 2B had 23 mM. The differences in sulfate concentrations could be explained by, for site 2A, a dilution of porewater with ascending fluids depleted in sulfate, or, the presence of gas hydrates supported by the observation of gas saturation. All this suggests that fluid and gas migration through the sedimentary layers is more active at site 2A than at site 2B.

Preliminary data of archaeal diversity was obtained by DNA extraction every 2 cm for each core, and the amplification of the 16S rRNA genes. The diversity was analyzed by DGGE. Bands were excised and sequenced, at the shallowest (0–2 cmbsf) and deepest (8–10 cmbsf) layers. Most sequences for both cores were affiliated with ANME archaea, involved in anaerobic methane oxidation. No differences were observed between both layers for each core. Pockmark sediments from site 2A were exclusively populated by ANME-2c archaea, whereas sediments from site 2B harboured a more diverse ANME community (ANME-1, -2a/b, and -2c). AOM does not appear to be the only microbial activity occurring in these sediments. Indeed, at both sites, sequences affiliated with the MBG-D archaea were found, which belong to the order *Thermoplasmatales*. Metagenomic reconstruction of MBG-D archaea have predicted their ability to cope with fluctuating environments (Lazar et al. 2017). Furthermore, biochemical pathways involved in CO_2-fixation were detected in the MBG-D genomes as well. The concomitant presence of *Thermoplasmatales* and ANME-2 archaea was reported in Kuroshima Knoll sediments, where intense migration of CO_2-rich fluids was reported (Inagaki et al. 2004), as well as in the water column of the Black Sea (Vetriani et al. 2003), and in the center of the Napoli mud volcano (Lazar et al. 2011a).

The data gathered for the Nile deep sea fan sediments highlight the ecological importance of the archaeal communities involved in AOM in pockmarks. The carbonate crusts observed on the seafloor may act as seals for the ascending methane, thus increasing methane availability in the analysed sediments layers. Therefore, cold seep ecosystems such as pockmarks, where gas and fluid emission are generally intense, could be privileged niches for the development of diverse ANME archaea populations, as observed in this survey of Mediterranean mud volcano sediments.

References

Brown HE, Holbrook WS, Hornbach MJ, Nealon J (2006) Slide structure and role of gas hydrate at the northern boundary of the Storegga slide, offshore Norway. Mar Geol 229:179–186

Charlou JL, Donval JP, Zitter T, Roy N, Jean-Baptiste JP, Foucher JP, Woodside J, MEDINAUT Scientific Party (2003) Evidence of methane venting and geochemistry of brines on mud volcanoes of the eastern Mediterranean Sea. Deep-Sea Res I 50:941–958

Corliss JB, Dymond J, Gordon LI, Edmond JM, von Herzen RP, Ballard RD et al (1979) Submarine thermal springs on the Galapagos Rift. Science 203:1073–1083

Daffonchio D, Borin S, Brusa T, Brusetti L, van der Wielen PWJJ, Bolhuis H, Yakimov MM, D'Auria G et al (2006) Stratified prokaryote network in the oxic–anoxic transition of a deep-sea halocline. Nature 440:203–207

Dählmann A, de Lange GJ (2003) Fluid-sediment interactions at Eastern Mediterranean mud volcanoes: a stable isotope study from ODP Leg 160. Earth Planet Sci Lett 212:377–391

Dimitrov L, Woodside J (2003) Deep sea pockmark environments in the eastern Mediterranean. Marine Geol 195:263–276

Dupré S, Woodside J, Foucher J-P, Lange G, Mascle J, Boetius A et al (2007) Seafloor geological studies above active gas chimneys off Egypt (Central Nile Deep Sea Fan). Deep Sea Res I 54:1146–1172

Fang J, Shizuka A, Kato C, Schouten S (2006) Microbial diversity of cold-seep sediments in Sagami Bay, Japan, as determined by 16S rRNA gene and lipid analyses. FEMS Microbiol Ecol 57:429–441

Girguis PR, Cozen AE, DeLong EF (2005) Growth and population dynamics of anaerobic methane-oxidizing archaea and sulfate-reducing bacteria in a continuous-flow bioreactor. Appl Environ Microbiol 71:3725–3733

Glissmann K, Chin K-J, Casper P, Conrad R (2004) Methanogenic pathway and archaeal community structure in the sediment of eutrophic lake Dagow: effect of temperature. Microb Ecol 48:389–399

Goffredi SK, Wilpiszeki R, Lee R, Orphan VJ (2008) Temporal evolution of methane cycling and phylogenetic diversity of archaea in sediments from a deep-sea whale-fall in Monterey Canyon, California. ISME J 2:204–220

Gontharet S, Pierre C, Blanc-Valleron M-M, Rouchy JM, Fouquet Y, Bayon G et al (2007) Nature and origin of diagenetic carbonate crusts and concretions from mud volcanoes and pockmarks of the Nile deep-sea fan (eastern Mediterranean Sea). Deep Sea Res II 54:1292–1311

Grünke S, Felden J, Lichtschlag A, Girnth A-C, De Beer D, Wenzhöfer F, Boetius A (2011) Niche differentiation among mat-forming, sulfide-oxidizing bacteria at cold seeps of the Nile Deep Sea Fan (Eastern Mediterranean Sea). Geobiology 9:330–348

Haese RR, Hensen C, de Lange GJ (2006) Pore water geochemistry of eastern Mediterranean mud volcanoes: implications for fluid transport and fluid origin. Mar Geol 225:191–208

Hallam SJ, Girguis PR, Preston CM, Richardson PM, DeLong EF (2003) Identification of methyl coenzyme M reductase A (mcrA) genes associated with methane-oxidizing Archaea. Appl Environ Microbiol 69:5483–5491

Heijs SK, Sinninghe-Damsté JS, Forney LJ (2005) Characterization of a deep-sea microbial mat from an active cold seep at the Milano mud volcano in the Eastern Mediterranean Sea. FEMS Microbiol Ecol 54:47–56

Heuer VB, Pohlman JW, Torres ME, Elvert M, Hinrichs K-U (2009) The stable carbon isotope biogeochemistry of acetate and other dissolved carbon species in deep subseafloor sediments at the northern Cascadia Margin. Geochim Cosmochim Acta 73:3323–3336

Holmer M, Kristensen E (1994) Coexistence of sulfate reduction and methane production in an organic-rich sediment. Mar Ecol Progr Ser 107:177–184

Inagaki F, Tsunogai U, Suzuki M, Kosaka A, Machiyama H, Takai K et al (2004) Characterization of C1-metabolizing prokaryotic communities in methane seep habitats at the Kuroshima Knoll, southern Ryukyu Arc, by analyzing pmoA, mmoX, mxaF, mcrA, and 16S rRNA genes. Appl Environ Microbiol 70:7445–7455

Jørgensen BB, Kasten S (2006) Sulfur cycling and methane oxidation. In: Schulz HD, Zabel M (eds) Marine Geochemistry, 2nd edn. Springer, New York, pp 271–309

Joye S, Connel TL, Miller LG, Oremland RS, Jellison RS (1999) Oxidation of ammonia and methane in an alkaline, saline lake. Limnol Oceanogr 44:178–188

Joye SB, Samarkin VA, Orcutt BN, MacDonald IR, Hinrichs KU, Elvert M, Teske AP, Lloyd KG et al (2009) Metabolic variability in seafloor brines revealed by carbon and sulphur dynamics. Nat Geosci 2:349–354

Knittel K, Lösekann T, Boetius A, Kort R, Amann R (2005) Diversity and distribution of methanotrophic Archaea at cold seeps. Appl Environ Microbiol 71:467–479

Knittel K, Boetius A (2009) Anaerobic oxidation of methane: progress with an unknown process. Ann Rev Microbiol 63:311–334

Kvenvolden KA (1995) A review of the geochemistry of methane in natural gas hydrate. Org Geochem 23:997–1008

Lazar CS (2010) Diversité et activité des communautés microbiennes dans des sédiments marins associés aux émissions de fluides froids. Ph.D. thesis, Université de Bretagne Occidentale, France

Lazar CS, L'Haridon S, Pignet P, Toffin L (2011a) Archaeal populations in hypersaline sediments underlying orange microbial mats in the Napoli mud volcano. Appl Environ Microbiol 77:3120–3131

Lazar CS, Parkes RJ, Cragg BA, L'Haridon S, Toffin L (2011b) Methanogenic diversity and activity in hypersaline sediments of the centre of the Napoli mud volcano, Eastern Mediterranean Sea. Environ Microbiol 13:2078–2091

Lazar CS, Parkes RJ, Cragg BA, L'Haridon S, Toffin L (2012) Methanogenic activity and diversity in the centre of the Amsterdam Mud Volcano, Eastern Mediterranean Sea. FEMS Microbiol Ecol 81:1–12

Lazar CS, Baker BJ, Seite KW, Teske AP (2017) Genomic reconstruction of multiple lineages of uncultured benthic archaea suggests distinct biogeochemical roles and ecological niches. ISME J 11:1–12

Levin LA (2005) Ecology of cold seep sediments: interactions of fauna with flow, chemistry and microbes. Oceanogr Marine Biol 43:1–46

Lloyd KG, Lapham L, Teske A (2006) An anaerobic methane-oxidizing community of ANME-1b archaea in hypersaline Gulf of Mexico sediments. Appl Environ Microbiol 72:7218–7230

Lykousis V, Alexandri S, Woodside J, de Lange G, Dählmann A, Perissoratis C, Heeschen K et al (2009) Mud volcanoes and gas hydrates in the Anaximander mountains (Eastern Mediterranean Sea). Mar Petrol Geol 26:854–872

Mills HJ, Hodges C, Wilson K, MacDonald IR, Sobecky PA (2003) Microbial diversity in sediments associated with surface-breaching gas hydrate mounds in the Gulf of Mexico. FEMS Microbiol Ecol 46:39–52

Mills HJ, Martinez RJ, Story S, Sobecky PA (2004) Identification of members of the metabolically active microbial populations associated with *Beggiatoa* species mat communities form Gulf of Mexico cold-seep sediments. Appl Environ Microbiol 70:5447–5458

Olu-Le-Roy K, Sibuet M, Fiala-Médioni A, Gofas S, Salas C, Mariotti A, Foucher JP, Woodside J (2004) Cold seep communities in the deep eastern Mediterranean Sea: composition, symbiosis and spatial distribution on mud volcanoes. Deep-Sea Res I 51:1915–1936

Omoregie EO, Mastalerz V, de Lange G, Straub KL, Kappler A, Røy H et al (2008) Biogeochemistry and community composition of iron- and sulfur-precipitating microbial mats at the Chefren mud volcano (Nile Deep Sea Fan, Eastern Mediterranean). Appl Environ Microbiol 74:3198–3215

Oren A (1999) Bioenergetic aspects of halophilism. Microbiol Mol Biol Rev 63:334–348

Paull CK, Hecker B, Commeau R, Freeman-Lynde RP, Neumann C, Corso WP et al (1984) Biological communities at the Florida escarpment resemble hydrothermal vent taxa. Science 226:965–967

Rossel PE, Elvert M, Ramette A, Boetius A, Hinrichs K-U (2011) Factors controlling the distribution of anaerobic methanotrophic communities in marine environments: evidence from intact polar membrane lipids. Geochim Cosmochim Acta 75:164–184

Roussel E, Sauvadet AL, Allard J, Chaduteau C, Richard P, Cambon-Bonavita MA, Chaumillon E (2009) Archaeal methane cycling communities associated with gassy subsurface sediments of Marennes-Oléron Bay (France). Geomicrobiol J 26:31–43

Sahling H, Rickert D, Lee RW, Linke P, Suess E (2002) Macrofaunal community structure and sulfide flux at gas hydrate deposits from the Cascadia convergent margin, NE Pacific. Mar Ecol Progr Ser 231:121–138

Suess E, Ritger SD, Moore JC, Jones ML, Kulm LD, Cochrane GR (1985) Biological communities at vent sites along the subduction zone off Oregon. Bull Biol Soc Washington 6:475–484

Vetriani C, Tran HV, Kerkhof LJ (2003) Fingerprinting microbial assemblages from the oxic/anoxic chemocline of the Black Sea. Appl Environ Microbiol 69:6481–6488

Wegener G, Shovitri M, Knittel K, Niemann H, Hovland M, Boetius A (2008) Biogeochemical processes and microbial diversity of the Gullfaks and Tommeliten methane seeps (Northern North Sea). Biogeosciences 5:1127–1144

Wellsbury P, Goodman K, Barth T, Cragg BA, Barnes SP, Parkes RJ (1997) Deep marine biosphere fuelled by increasing organic matter availability during burial and heating. Nature 388:573–576

Zitter TAC, Huguen C, Woodside JM (2005) Geology of mud volcanoes in the eastern Mediterranean from combined sidescan sonar and submersible surveys. Deep-Sea Res I 52:457–475

Chapter 7
The Microbial Communities of the East Mediterranean Sea Mud Volcanoes and Pockmarks

Konstantinos Ar. Kormas and Alexandra Meziti

Abstract Mud volcanoes and pockmarks are sites which methane and other gases seep out towards the overlying water. This supply and the microbiologically-mediated processes of these cold seeps can be the basis of unique microbial habitats. Although mud volcanoes and pockmarks vary in the seepage rate and chemical composition of their outflows, the communities found in geographically adjacent systems seem to harbor comparable microbial communities. For the East Mediterranean Sea, only sporadic data exist on its major mud volcanoes and pockmarks collected through a few oceanographic cruises; however, further the knowledge of these systems should be gained by microbiological and sequence-based investigations.

Introduction

Mud volcanoes and pockmarks are among the major types of cold seeps found in the deep-sea (Boetius and Wenzhöfer 2013). Due to their very specific environmental features, which are very distinct from the surrounding deep-sea sediments (Vanreusel et al. 2009), they can be considered as microbial hot-spot habitats, whose basal functioning relies solely on their residing microorganisms (see also Chap. 6). Almost a decade after the discovery of the Mediterranean Sea methane-dominated deep-sea habitats (Cita et al. 1981) environmental microbiology was entering its molecular era (Amann et al. 1995), giving unprecedented capabilities for detailed studies of microbial communities in practically every natural sample. This opportunity led to the first molecular-microbiological studies of the Mediterranean Sea deep-sea mud volcanoes in the late 1990s.

K. Ar. Kormas (✉) · A. Meziti
University of Thessaly, Volos, Greece
e-mail: kkormas@uth.gr

A. Meziti
e-mail: ameziti@uth.gr

A. Teske and V. Carvalho (eds.), *Marine Hydrocarbon Seeps*,
Springer Oceanography, https://doi.org/10.1007/978-3-030-34827-4_7

The microbial diversity of marine mud volcanoes includes prokaryotic communities of specific structure and ecophysiological roles, related mostly to the anaerobic oxidation of methane (Knittel and Boetius 2009), although this ecologically important process can be associated with other electron acceptors than sulfate (Scheller et al. 2016). As prokaryotic diversity seems to exhibit biogeographical patterns even at the scale of large bathypelagic regions (Amend et al. 2013; Salazar et al. 2016) and pelagic sediments (Bienhold et al. 2016) it is expected that more specialized habitats, such as marine mud volcanoes and cold seeps, should hold more specialized microbial communities. Indeed, Pachiadaki and Kormas (2013) showed that among five methane seep-dominated habitats of the European continental shelf there is some degree of inter-connectivity of their prokaryotic communities, attributed mostly to consortia of methane-oxidizing archaea and sulfate-reducing bacteria. However, while these fairly well studied microbial consortia often dominate mud volcanoes, these habitats also yield unusual microbial associations, such as the recent discovery of ANME-associated *Verrucomicrobia* (Hatzenpichler et al. 2016). Apart from their rather well-known microorganisms, marine mud volcanoes seem to harbor truly novel microorganisms as well. Initial studies of the Mediterranean mud volcanoes have revealed a significant number of yet-uncultivated, phylum-level lineages, such as JS1, WS3, OP8, and OP11. Currently, some of the genomes of selected representatives along with their potential metabolic repertoire are known (e.g. Rinke et al. 2013; Solden et al. 2016).

The Eastern Mediterranean Sea Mud Volcanoes and Pockmarks: Sites and Methods Used

The available scientific literature on the microbial diversity of the Eastern Mediterranean Sea mud volcanoes, pockmarks, and hydrocarbon seeps is rather limited. So far, six mud volcanoes and cold seeps of the eastern Mediterranean area have been studied in terms of molecular phylogenetic diversity of *Bacteria* and *Archaea* (on the latter, see Chap. 7). More specifically, phylogenetic studies have been performed at the mud volcanoes Amsterdam and Kazan from the Anaximander mountains (Heijs et al. 2007; Kormas et al. 2008; Pachiadaki et al. 2010, 2011), Milano and Napoli from the Olimpi mountains (Heijs et al. 2005), at the cold seeps Isis, Amon, and Chefren from the Nile Deep-Sea Fan (NDSF) (Omoregie et al. 2008, 2009), and at two hydrocarbon-related sites in the Levantine Basin (Rubin-Blum et al. 2014).

Most studies (Kormas et al. 2008; Heijs et al. 2007; Pachiadaki et al. 2010, 2011; Rubin-Blum et al. 2014) included phylogenetic analysis of the 16S rRNA gene diversity in various sediment layers, while the rest studied diversity of surface sediment samples (Omoregie et al. 2008, 2009), or microbial mats (Heijs et al. 2008; Omoregie et al. 2008). All studies used universal (but different pairs of) bacterial and archaeal primers and studied diversity with the use of clone libraries, while some studies (Omoregie et al. 2008, 2009) used Fluorescence in situ Hybridization (FISH) as well. In all studies, phylogenetic data were combined with physicochemical properties, including sulfate, methane and hydrogen sulfide concentrations in the studied sites.

Bacterial Diversity

All sites exhibited high relative abundances of *Deltaproteobacteria*, and more specifically, enhanced presence (>30%) of sulfate-reducing bacteria (SRB) of the *Desulfosarcina, Desulfococcus, Desulfobacter*, and *Desulfobulbus* clusters. SRB representatives occurred in high abundances in samples where anaerobic oxidation of methane (AOM) was either inferred (Kormas et al. 2008; Pachiadaki et al. 2010, 2011; Rubin-Blum et al. 2014) or measured (Omoregie et al. 2008, 2009), and in depths down to 14 cmbsf. The presence of the SRB group was coupled with increasing sulfide concentration, with the exception of Amsterdam MV (Heijs et al. 2008), where increased *Deltaproteobacteria* abundances were concomitantly measured with increased sulfate concentrations, implying either the advective fluid flow or the enhanced mixing of the specific samples.

Members of *Actinobacteria* were observed in significant abundance in the surface samples of all studies that included also deeper sediment layers (Heijs et al. 2008; Pachiadaki et al. 2010, 2011), and their presence in mud volcano sediments could be attributed to the use of acetate that is released in environments were sulfate reduction takes place. In the Levantine Basin seeps, however, such *Actinobacteria* that were related to fermenting microorganisms, were absent and were found only in the control sediment samples. *Actinobacteria* could also be involved in the degradation of complex organic molecules such as lignin (Stach and Bull 2005), so it can be hypothesized that they could also be key players for the degradation and the remineralization of complex organic matter at mud volcano sites. *Chloroflexi* representatives appeared in the same studies, but they were significantly increased in deeper anoxic layers.

Gammaproteobacteria were present in most studied sites while no specific pattern was observed. At the microbial mats of the Chefren mud volcano (Omoregie et al. 2009) *Gammaproteobacteria* were dominant in the bacterial population of the microbial mats as well as in the underlying sediments, under them, with representatives clustering mainly in the methanotrophs. At Kazan mud volcano (Pachiadaki et al. 2010) they dominated at the surface and deeper layers (25–30 cmbsf), while at the Amsterdam mud volcano (Pachiadaki et al. 2011) they dominated all surface samples. In this case, the dominance of *Gammaproteobacteria* was attributed to their opportunistic life strategies. The phylotypes observed in Kazan and Amsterdam mud volcanoes were mostly related with sulfur and sulfide-oxidizers, and were detected in depths where sulfur and sulfide oxidation could potentially occur. However, the *Gammaproteobacteria* did not reflect any specific occurrence patterns like the SRB.

The overall dominance of *Deltaproteobacteria*, *Epsilonproteobacteria*, and *Gammaproteobacteria* was also shown in a spatially detailed study in nine methane-related habitats, i.e. Amon and Amsterdam mud volcanoes, as well as pockmarks of the Nile Fan (Pop-Ristova et al. 2015).

The highest diversity in terms of the presence of different bacterial groups was observed in Kazan and Amsterdam mud volcanoes in the studies performed by Pachiadaki et al. (2010, 2011), possibly due to the more contemporary and thorough analysis performed. *Acidobacteria, Bacteroidetes, Chlorobi, Defferibacteres, Epsilonproteobacteria, Firmicutes, Planctomycetes, Spirochaetes, Verrucomicrobia*, and the candidate divisions JS1, OD1/OP11, OP8, WS1, WS3, WS6, and OP5 were the groups that were detected occasionally in all mud volcano sites without exhibiting any pattern or dominance over the above mentioned bacterial groups.

Archaeal Diversity

In all currently available studies, representatives of the anaerobic methanotrophs (ANME) group were detected (see also Chap. 6). In most cases, ANME representatives were found together with bacteria of the SRB group, providing evidence for the anaerobic oxidation of methane (AOM). The mats at the mud volcanoes of the Nile Deep Sea Fan (NDSF) were mostly dominated by ANME-2 representatives, as shown through clone libraries and FISH (Omoregie et al. 2008, 2009). FISH showed that ANME-2 archaea, present in Amon and Chefren created aggregates with SRB, while increased abundance of ANME-2 in clone libraries and FISH counts matched with the high AOM rates measured (Omoregie et al. 2009). Other archaeal groups in microbial mats of the NDSF were ANME-3 and MBGD-D, while the presence of *Crenarchaeota* groups (MBGB, MBG-1) was limited. The importance of the ANMEs in hydrocarbon seeps of the Eastern Mediterranean was also indicated by the occurrence of ANME-2c, ANME-3, and two novel potential groups related to the ANME in two hydrocarbon-related deep-sea sites of the Eastern Mediterranean Levantine Basin (Rubin-Blum et al. 2014).

Similarly, at Kazan mud volcano, representatives of ANME-2 were dominant in deeper layers, where AOM was implied to occur on the basis of physicochemical data (Kormas et al. 2008; Pachiadaki et al. 2010). In the study of Kormas et al. (2008) the dominance of ANME-2 was accompanied by the detection of *mcrA* genes that also clustered among those of other ANME-2. In Kazan mud volcano, also other groups of ANME were detected (Pachiadaki et al. 2010), with ANME-3 dominating close to the surface and ANME-1 at greater depths. The study of Heijs et al. (2008) at Kazan mud volcano showed co-occurrence of ANME-1 and ANME-2, although ANME-2 representatives were more prevalent. ANME groups are the key players in AOM in marine sediments (Knittel and Boetius 2009; and Chaps. 1 and 4).

The high sediment depth resolution study of Amsterdam mud volcano by Pachiadaki et al. (2011) revealed a spatial differentiation of the dominance patterns of ANMEs. ANME-2b dominated the surface layer and they were gradually replaced by ANME-1 at 10 cmbsf, ANME-2a/c at 20 cmbsf, and finally by ANME-3 at 30 cmbsf. A similar pattern was not observed in the study of Heijs et al. (2008) at Amsterdam mud volcano, that instead detected a dominance of ANME-1-related *Archaea* at all depths investigated, but this could be also due to the lower depth resolution in that study.

Future Perspectives

The Mediterranean Sea mud volcanoes and pockmarks have been sporadically investigated. The available studies to date include only prokaryotic diversity and ecology, with no available data on microscopic eukaryotes, namely fungi and protists, of these habitats. Direct comparisons of whole communities between the different sites are challenging due to several issues, including technical practicalities (e.g. different sampling and DNA extraction protocols, different target regions of the 16S rRNA gene). Another difficulty for comparison of the sites is the lack of a focused temporal study on the changes of these microbial communities, and how they respond to the stable versus fluctuating geological and chemical conditions of the mud volcanoes, pockmarks and hydrocarbon seeps. Pachiadaki and Kormas (2013) were able to depict the most important organisms in these systems, at least in terms of presence/absence, by comparing the occurrence of shared archaeal and bacterial phylotypes occurring in different mud volcanoes to European continental shelf sediments. An expansion of such studies, including all related marine habitats, along with the inclusion of temporal changes of their microbial communities and the latest technologies, such as single cell genomics and meta-omics approaches, would elucidate the degree of distinctiveness of the mud volcanoes habitats and their ecophysiological roles imposed by their microbial communities.

References

Amann R, Ludwig W, Schleifer K (1995) Phylogenetic identification and in situ detection of individual microbial cells without cultivation. Microbiol Rev 59:143–169

Amend AS, Oliver TA, Amaral-Zettler LA, Boetius A et al (2013) Macroecological patterns of marine bacteria on a global scale. J Biogeogr 40:800–811

Bienhold C, Zinger L, Boetius A, Ramette A (2016) Diversity and biogeography of bathyal and abyssal seafloor bacteria. PLoS ONE 11:e0148016

Boetius A, Wenzhöfer F (2013) Seafloor oxygen consumption fueled by methane from cold seeps. Nat Geosci 6:725–734

Cita MB, Ryan WBF, Paggi L (1981) Prometheus mud-breccia: An example of shale diapirism in the western Mediterranean ridge. Ann Geol Pays Hellen 30:543–570

Hatzenpichler R, Connon SA, Goudeau D, Malmstrom RR, Woyke T, Orphan VJ (2016) Visualizing in situ translational activity for identifying and sorting slow-growing archaeal—bacterial consortia. Proc Natl Acad Sci USA 113:E4069–E4078

Heijs SK, Haese RR, Van Der Wielen PWJJ, Forney LJ, Van Elsas JD (2007) Use of 16S rRNA gene based clone libraries to assess microbial communities potentially involved in anaerobic methane oxidation in a Mediterranean cold seep. Microb Ecol 53:384–398

Heijs SK, Laverman AM, Forney LJ, Hardoim PR, Van Elsas JD (2008) Comparison of deep-sea sediment microbial communities in the eastern Mediterranean. FEMS Microbiol Ecol 64:362–377

Heijs SK, Sinninghe Damsté JS, Forney LJ (2005) Characterization of a deep-sea microbial mat from an active cold seep at the Milano mud volcano in the eastern Mediterranean Sea. FEMS Microbiol Ecol 54:47–56

Knittel K, Boetius A (2009) Anaerobic oxidation of methane: progress with an unknown process. Ann Rev Microbiol 63:311–334

Kormas KA, Meziti A, Dahlmann A, De Lange GJ, Lykousis V (2008) Characterization of methanogenic and prokaryotic assemblages based on mcrA and 16S rRNA gene diversity in sediments of the Kazan mud volcano (Mediterranean Sea). Geobiology 6:450–460

Omoregie EO, Mastalerz V, de Lange G, Straub KL et al (2008) Biogeochemistry and community composition of iron- and sulfur-precipitating microbial mats at the Chefren mud volcano (Nile Deep Sea Fan, eastern Mediterranean). Appl Environ Microbiol 74:3198–3215

Omoregie EO, Niemann H, Mastalerz V, de Lange GJ et al (2009) Microbial methane oxidation and sulfate reduction at cold seeps of the deep eastern Mediterranean Sea. Mar Geol 261:114–127

Pachiadaki MG, Kallionaki A, Shlmann D, De Lange A, Kormas KJ (2011) Diversity and spatial distribution of prokaryotic communities along a sediment vertical profile of a deep-sea mud volcano. Microb Ecol 62:655–668

Pachiadaki MG, Kormas KA (2013) Interconnectivity versus isolation of prokaryotic communities in European deep-sea mud volcanoes. Biogeosciences 10:2821–2831

Pachiadaki MG, Lykousis V, Stefanou EG, Kormas KA (2010) Prokaryotic community structure and diversity in the sediments of an active submarine mud volcano (Kazan mud volcano, East Mediterranean Sea). FEMS Microbiol Ecol 72:429–444

Pop Ristova P, Wenzhofer F, Ramette A, Felden J, Boetius A (2015) Spatial scales of bacterial community diversity at cold seeps (eastern Mediterranean Sea). ISME J 9:1306–1318

Rinke C, Schwientek P, Sczyrba A, Ivanova NN et al (2013) Insights into the phylogeny and coding potential of microbial dark matter. Nature 499:431–437

Rubin-Blum M, Antler G, Turchyn AV, Tsadok R et al (2014) Hydrocarbon-related microbial processes in the deep sediments of the eastern Mediterranean Levantine Basin. FEMS Microbiol Ecol 87:780–796

Salazar G, Cornejo-Castillo FM, Benitez-Barrios V, Fraile-Nuez E et al (2016) Global diversity and biogeography of deep-sea pelagic prokaryotes. ISME J 10:596–608

Scheller S, Yu H, Chadwick GL, McGlynn SE, Orphan VJ (2016) Artificial electron acceptors decouple archaeal methane oxidation from sulfate reduction. Science 351:703–707

Solden L, Lloyd K, Wrighton K (2016) The bright side of microbial dark matter: Lessons learned from the uncultivated majority. Curr Opin Microbiol 31:217–226

Stach EM, Bull A (2005) Estimating and comparing the diversity of marine Actinobacteria. Antonie Van Leeuwenhoek 87:3–9

Vanreusel A, Andersen AC, Boetius A, Connelly D et al (2009) Biodiversity of cold seep ecosystems along the European margins. Oceanography 22:110–127

Chapter 8
Large Sulfur-Oxidizing Bacteria at Gulf of Mexico Hydrocarbon Seeps

Andreas Teske and Verena Carvalho

Abstract Large sulfide-oxidizing bacteria occur as visually conspicuous microbial mats in a wide range of sedimentary habitats, including estuarine and coastal marine sediments, and deep-sea vents and seeps. The microbial mats spread on the surface of sulfide-rich sediments, thus intercepting and oxidizing sulfide that diffuses upwards from the underlying sulfate reduction zone, or, that reaches the surface by advection of reduced fluids. The first intermediate of bacterial sulfide oxidation, elemental sulfur, is stored within the cytoplasm as globules, and serves as energy reserve. Large sulfide oxidizers have a wide metabolic repertoire, including autotrophic carbon fixation, sulfide and sulfur oxidation to sulfuric acid, nitrate reduction to ammonia or nitrogen gas, as well as polyphosphate storage and release causing local phosphate supersaturation and precipitation. Large sulfur-oxidizing bacteria are widespread at hydrocarbon seeps in the Gulf of Mexico, where numerous types with different morphology, phylogenetic affiliation, and physiology have been documented. In this chapter, we provide an overview of large sulfur-oxidizing bacteria in the Gulf of Mexico. We also incorporate previously unpublished sequencing data for selected filaments, and include recent observations of new morphological variants, including one that resembles sheathed marine *Thioploca* spp., but which shows a distinguished, unique branching morphology.

Initial Surveys

Initial studies of large sulfur-oxidizing bacteria in the Gulf of Mexico described white and orange mats of large filamentous bacteria growing on seep sediments and hydrate mounds in the Green Canyon area (MacDonald et al. 1989; Roberts et al.

A. Teske (✉)
University of North Carolina at Chapel Hill, Chapel Hill, NC, USA
e-mail: teske@email.unc.edu

V. Carvalho
Max Planck Institute for Marine Microbiology, Bremen, Germany
e-mail: vcarvalh@mpi-bremen.de

A. Teske and V. Carvalho (eds.), *Marine Hydrocarbon Seeps*,
Springer Oceanography, https://doi.org/10.1007/978-3-030-34827-4_8

1990; Sassen et al. 1993; see also Chap. 4) While the majority of mat biomass appeared white or colorless, some areas were dominated by denser and orange-colored mats, which were often co-localized with oily sediments (MacDonald et al. 1989). The largest filaments showed diameters of ca. 200 μm and were several cm long (Larkin and Henk 1996). Light microscopy revealed that the majority of organisms forming the mat biomass are filamentous and contain small inclusions, presumably sulfur granules. A detailed electron microscopic study of these filaments showed that they are hollow (Larkin and Henk 1996), confirming observations of large filamentous sulfur-oxidizing bacteria from other deep-sea sediments and vent sites. This "hollow" morphology of a large central/aqueous vacuole confining the cytoplasm to a thin layer along the cell walls was soon recognized as a typical feature of large, sulfur-oxidizing bacteria (Schulz and Jørgensen 2001).

Sediments in the Gulf of Mexico containing these large conspicuous microbial mats were found over a wide depth range, from the photic zone at 130 m and down the slope to depths around 1000 m (e.g., Larkin and Henk 1996; Salman-Carvalho et al. 2016; Sassen et al. 1993). Filamentous microbial mats show a generally wider environmental distribution than chemosynthetic seep fauna, and are commonly associated with sea floor features related to hydrocarbon seepage, such as small sediment fractures, steeply-dipping fault scarps in sediments, infaunal burrows, pockmarks or craters, mud and brine flow, gas hydrate mounds, and fault vents in carbonate outcrops (Sassen et al. 1993). Some mats were even observed to cover the fluid mud surface within the crater of an active mud vent in Green Canyon (Roberts and Neurauter 1990). The light $\delta^{13}C$ isotopic value of a biomass sample consisting of large filaments (−27.9‰) was originally interpreted as evidence for chemosynthetic activity, assimilating CO_2 derived from bacterial degradation of hydrocarbons (Sassen et al. 1993). Thus, the ecological and biogeochemical function of the mat-forming filaments was hypothesized to link sulfide and sulfur oxidation to carbon fixation, in particular CO_2 remobilized from microbial hydrocarbon oxidation in the sediment (Sassen et al. 1993). Subsequent $\delta^{13}C$ isotopic analyses of biomass from filamentous mats at diverse seep locations in the Gulf of Mexico resulted in a similar range (−26.6 to −28.6‰; Aharon and Fu 2000; Sassen et al. 1993; Zhang et al. 2005). However, the interpretation of these data has changed since then, given that heterotrophic uptake and assimilation of photosynthetically-derived biomass, or uptake of soluble substrates derived from microbial hydrocarbon degradation, could produce similar $\delta^{13}C$ signatures (Zhang et al. 2005). Investigating the question of heterotrophic versus autotrophic metabolism requires physiological tests of carbon uptake in living filaments.

During the initial time of discovery, filaments that formed the bulk biomass of the mats were identified as members of the genus *Beggiatoa* based on morphological resemblance. However, these filaments are unlikely to be related to the genus *Beggiatoa* in the strict sense, as it consists of non-vacuolated, thin filamentous bacteria from freshwater or brackish habitats. Also acknowledging the fact that mats do not consist of a single organism (species, phylotype, ecotype etc.) it is more appropriate to refer to the mat-forming sulfur-oxidizers as members of the family *Beggiatoaceae* (Salman et al. 2011) until sequence data for phylogenetic identification become available. Several studies sampled these Gulf of Mexico mats dominated by filaments that were morphologically reminiscent of "*Beggiatoa*", and sequenced their 16S rRNA genes (Lloyd et al. 2010; Mills et al. 2004; Salman-Carvalho et al. 2016; Zhang et al. 2005). Within the complex mat populations, the phylogenetic affiliation of selected sulfide-oxidizing filaments was very heterogeneous (Fig. 1), which confirms that large sulfide-oxidizing bacteria are not reliably identified by morphology alone (Salman et al. 2011, 2013).

Pigmented and Unpigmented Filamentous Beggiatoaceae

Filamentous sulfur bacteria constitute the main biomass of the extensive sediment-covering mats in the Gulf of Mexico. These filaments include non-pigmented (white) and pigmented (yellow to orange) forms, which grow in spatially distinct mat patches (Fig. 2; Larkin and Henk 1996; MacDonald et al. 1989; Mills et al. 2004; Nikolaus et al. 2003; Sassen et al. 1993; Sen Gupta et al. 1997). The unusual orange pigmentation is reminiscent of similarly large, pigmented filaments observed at cold seeps in Monterey Canyon (Barry et al. 1996), on top of hydrothermally influenced sediments at Endeavor Ridge (Hedrick et al. 1992), and in the Guaymas Basin (McKay et al. 2012). Orange-pigmented and unpigmented filaments collected at the Gulf of Mexico Bush Hill hydrate site (GC185) showed overlapping filament diameters. The orange types were mostly smaller, 25–35 μm in diameter, and ranged from <25–65 μm, while the unpigmented filaments exhibited a greater range towards larger filament sizes, <25 to >85 μm (Nikolaus et al. 2003). As shown for Guaymas Basin, where a smaller pigmented and a larger unpigmented filament type form two different 16S rRNA lineages within the *Beggiatoaceae* (McKay et al. 2012), and it can be assumed that these differently pigmented filament size classes in the Gulf of Mexico may represent physiologically and phylogenetically distinct taxa. The orange pigment of Gulf of Mexico filaments varied in intensity among and even within individual filaments. It was soluble in distilled water and in the aqueous assay buffer used for CO_2 assimilation tests, and had an absorbance peak near 390 nm (Nikolaus et al. 2003). The easy solubility suggests a cytoplasmatic, periplasmatic, or, loosely membrane-associated protein. Interestingly, the orange pigment studied in Guaymas Basin filaments, an octaheme cytochrome, was also easily soluble, and when partially purified and assayed for enzymatic activity it showed multifunctional activity as a nitrite reductase, hydroxylamine oxidoreductase, and hydrazine oxidase (MacGregor et al. 2013). Such enzymatic activity tests have not yet been performed for the pigmented filaments from the Gulf of Mexico, however, another interesting physiologically distinction could be determined between pigmented and unpigmented filaments. Shipboard experiments showed that the unpigmented were

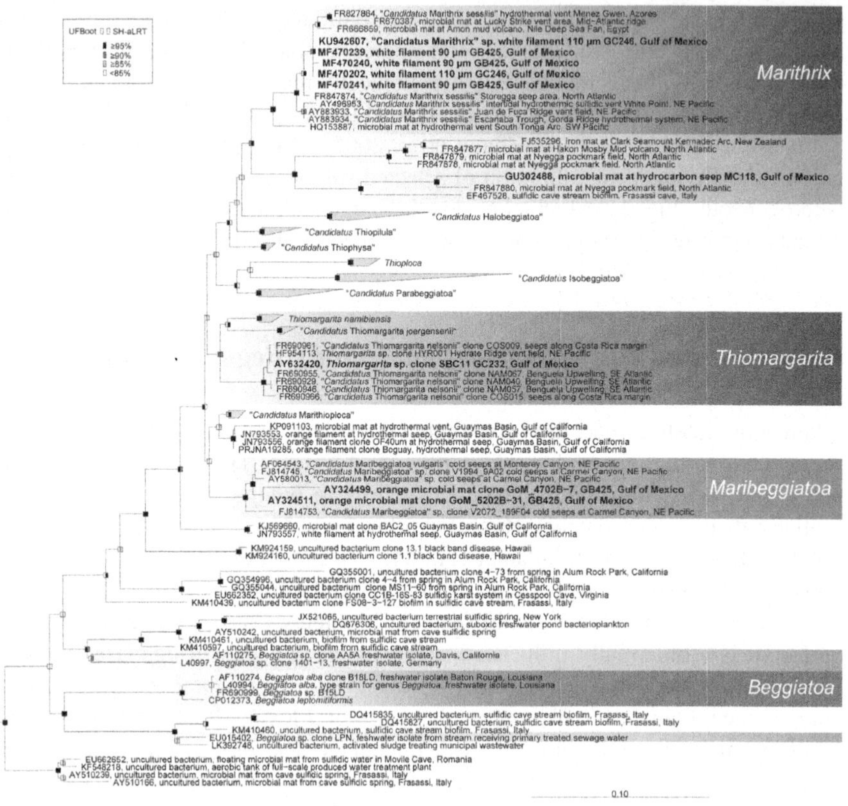

Fig. 1 Phylogenetic tree based on 16S rRNA gene sequences of the family *Beggiatoaceae* (*Gammaproteobacteria*). Highlighted in bold are those sequences originating from large sulfide-oxidizing bacteria from Gulf of Mexico sediments, and the genera they affiliate to are colored and labeled. As reference, also the original organisms classified as *Beggiatoa* are colored in green. For tree reconstruction, a total of 157 near full-length 16S rRNA gene sequences were selected from SILVA (Quast et al. 2013) and the public databases, and aligned in ARB with the SINA aligner (Ludwig et al. 2004). The phylogeny was then inferred using a maximum likelihood approach implemented in IQ Tree (Nguyen et al. 2014). Based on the Bayesian Information Criterion (BIC) using the model finder tool in IQ Tree (Kalyaanamoorthy et al. 2017) a TIM2 model including 4 rate categories of distribution-free rates was the best-fit nucleotide substitution model to the sequence alignment. Branch support was assessed using 1,000 replicates of ultrafast bootstrap (UFBoot; Minh et al. 2013) and the Shimodaira–Hasegawa approximate likelihood-ratio test (SH-aLRT; Guindon et al. 2010). After reimporting the tree back into ARB short sequences were added with parsimony criteria without changes in the tree backbone topology. The tree was rooted with representatives of the *Desulfobacteraceae* (*Deltaproteobacteria*)

capable of significant CO_2 fixation, while the orange filaments showed very little incorporation activity. The uptake rate of unpigmented filaments further increased with additions of thiosulfate and sulfide as electron donors (from 0.017 to 0.033 nmol CO_2 mg dry-weight^{-1} min^{-1}), while activity of pigmented filaments

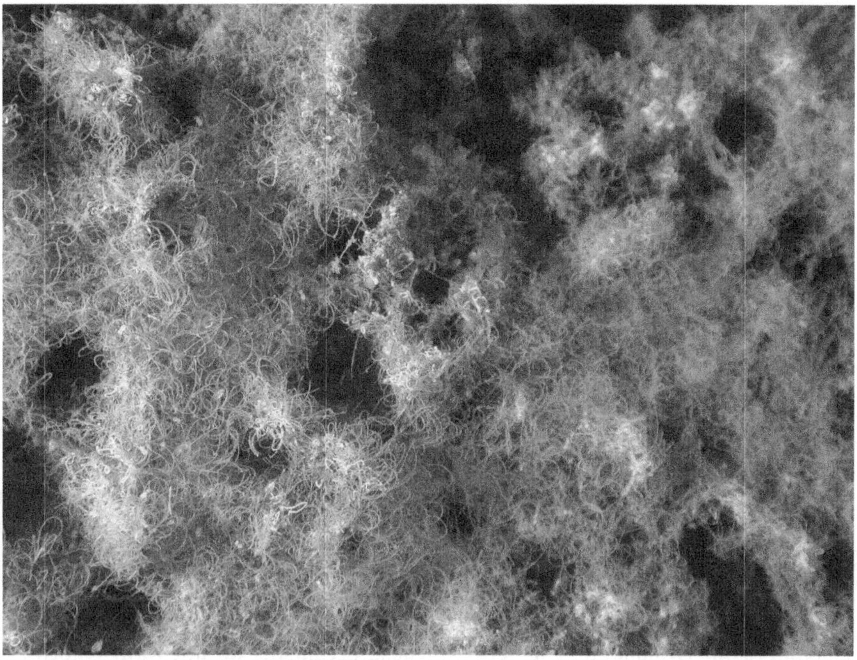

Fig. 2 In-situ close-up photograph of predominantly orange filamentous mats at Green Canyon 185, a hydrate-rich seep site in the Gulf of Mexico (N 27°44.930, W 91°30.450; 540 m depth). Within the complex mat architecture, individual spherical sulfur bacteria (most likely *Thiomargarita* spp.) occur amidst the dominant white and orange filaments growing in thick tufts. The photo was obtained with a custom-designed digital camera array mounted on the submersible arm (Nikolaus et al. 2003). The image was taken by I. R. MacDonald, Florida State University

remained around 0.004 nmol CO_2 mg dry-weight^{-1} min^{-1}. A RuBisCO assay at 15 °C, pH 8.5, and supplemented with 10 mM ribulose bisphosphate showed that the unpigmented filaments had ca. 20× higher activities than the orange filaments (9.65 vs. 0.41 nmol CO_2 fixed mg protein^{-1} min^{-1}). RuBisCO assays over a wider temperature and pH range showed variable, but generally higher values for unpigmented filaments (up to two orders of magnitude) than pigmented filaments (Nikolaus et al. 2003). These results are congruent with previous experiments published in an expedition report, where unpigmented Gulf of Mexico seep filaments were assayed at in situ temperature of 8 °C, and showed high CO_2 uptake rates averaging 57.5 nmol CO_2 mg dry-weight^{-1} h^{-1} (Wirsen et al. 1992). Activities of unpigmented mats were also stimulated by sulfide and thiosulfate additions whereas orange-pigmented filaments showed no CO_2 fixation activity (Wirsen et al. 1992). These consistent results indicate that unpigmented, large filaments may be chemolithoautotrophs that use CO_2 as a carbon source and H_2S as electron donor. In contrast, pigmented filaments may have a rather mixotrophic lifestyle using H_2S as an energy source while relying on organic carbon compounds for biosynthetic purposes, or, they may even be organoheterotrophs, using organic carbon both as energy and carbon source (Nikolaus et al. 2003). Further work with Gulf of Mexico mat-forming filaments is required to elucidate their metabolism in greater detail, combined with genomic surveys and better taxonomic resolution to reliably identify the predominant filament types.

Unicellular Thiomargarita

Besides observing mat-forming filaments, Larkin and Henk (1996) also found large single spherical sulfur-containing cells in the Gulf of Mexico sediments. Based on morphological resemblance they referred to these organisms as *Achromatium*, despite the fact that they could not observe the typical calcite inclusions of *Achromatium* spp. Upon handling, they observed these cells to collapse, indicative of the presence of a central aqueous vacuole. These features are reminiscent of *Thiomargarita*, a genus of non-filamentous vacuolated large sulfur bacteria that were discovered a few years later in sediments of the Benguela Upwelling region (Schulz et al. 1999). Subsequently, the large non-filamentous single cells in the Gulf of Mexico were identified phylogenetically as members of the genus *Thiomargarita* (Kalanetra et al. 2005). These organisms do not form colorful or white mats comparable to those of filamentous sulfur oxidizers, which makes them less conspicuous to casual observers. Yet, populations of large, nitrate-accumulating *Thiomargarita* cells were found in the majority of seep sediment cores taken near a brine pool (GC233), a mud volcano (GB425), and from sediment-covered hydrate mounds (GC234 and GC185) (Kalanetra et al. 2005). Upon close-up inspection of sediment cores in the shipboard lab, the *Thiomargarita* cells appeared as single, immotile, non-pigmented and frequently particle-encrusted cells on the sediment surface (Kalanetra et al. 2005). Close-up seafloor photography shows *Thiomargarita* cells associated with the complex, rough seafloor texture at seep sites that is characterized by many small cracks, fissures, tufts of filamentous bacteria, and olive-brown hydrocarbon stains (Fig. 3a). Their occurrence in the proximity of filamentous sulfur oxidizers is frequently observed, but apparently not obligatory, suggesting that *Thiomargarita* cells may be more widespread than the highly conspicuous mat-forming filaments.

Thiomargarita-harboring sediments are characterized by high rates of sulfate reduction fueled by hydrocarbon oxidation (Arvidson et al. 2004; Joye et al. 2004). Like *Thiomargarita* cells from the Benguela Upwelling system offshore Namibia, the Gulf of Mexico cells shared exceptionally large cell sizes (ca. 180–360 μm cell diameter), and featured a large vacuole that contained high concentrations of nitrate, and that reduced the cytoplasmic space to a thin layer along the cell wall (Kalanetra et al. 2005). The Namibian and Gulf of Mexico populations are phylogenetically related, and form a monophyletic branch among the large filamentous, vacuolated, nitrate-accumulating sulfur-oxidizing bacteria (Kalanetra et al. 2005; Fig. 1). New *Thiomargarita* sequences and sister lineages have been added since then, but they

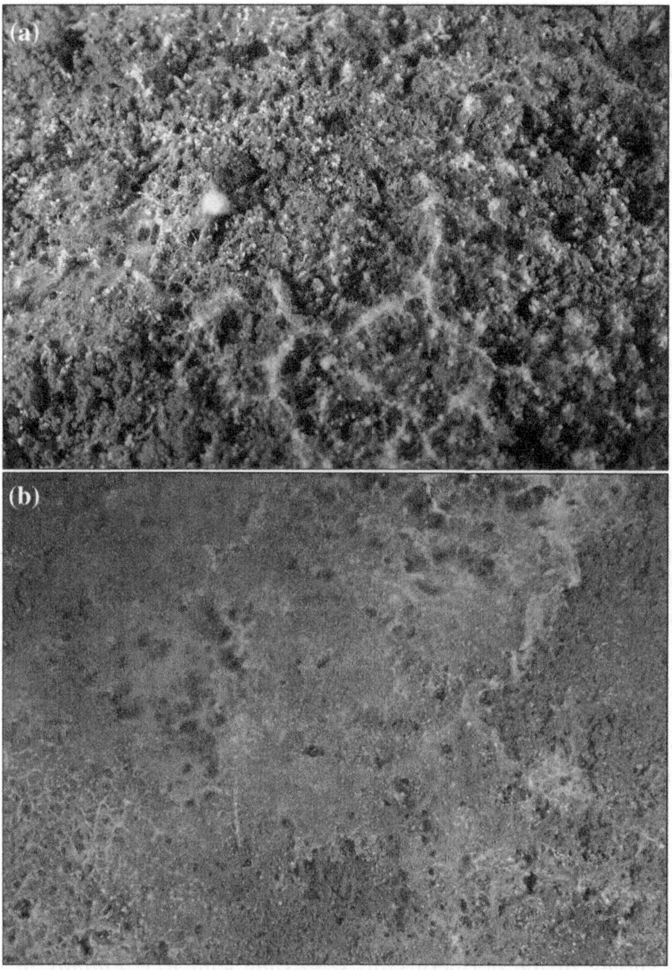

Fig. 3 Sediment with *Thiomargarita* cells at Bush Hill, GC 185. The *Thiomargarita* cells appear in small clusters or as single cells, occurring together with small tufts of unpigmented large filaments (**a**), both types are contrasting white against the darker, olive-grey sediment. Orange-pigmented filamentous mats, non-pigmented veil-like structures, and small white *Thiomargarita* cells are growing adjacent to each other on brown sediment (**b**). Both photos were obtained with a custom-designed digital camera array mounted on the submersible arm (Nikolaus et al. 2003) by I. R. MacDonald, Florida State University

all branch close to each other in the phylogenetic framework of the family *Beggiatoaceae* (Salman et al. 2011; Teske and Salman 2014). In contrast to the Namibian *Thiomargarita* cells that typically form chains surrounded by a shared sheath, and grow by cell division along the chain axis, the Gulf of Mexico *Thiomargarita* grew as individual cells that underwent cell division in three dimensions, forming duplets, triplets and quartets in one, two, and three division steps (Kalanetra et al. 2005). These multi-planar cell divisions were frequently observed in *Thiomargarita*-rich sediment samples that had been stored in the lab for 20 months (between July 2002 and February 2004) after the sampling expedition. Cell division went hand in hand with a reduction in cell diameter (on average from 262 to 163 μm), and a reduction in cell volume by two or three orders of magnitude (Kalanetra et al. 2005). This diminishment of cell size and cell volume could represent a starvation response, related to gradually decreasing sulfide flux in the stored sediment cores that lack a fresh supply of hydrocarbons. These peculiar reductive division stages have also been observed in both, stored and fresh sediment samples from the Benguela Upwelling zone (Salman et al. 2013).

At present, the conditions and geochemical controls that favor either filamentous *Beggiatoaceae* or non-filamentous, unicellular *Thiomargarita* cells at Gulf of Mexico seeps are not well known. The wide distribution of *Thiomargarita* cells in Gulf of Mexico sediments, and their close relationship with Namibian *Thiomargarita* suggests a greater degree of tolerance to different, possibly fluctuating seepage conditions, including greater perseverance under starvation conditions as shown by extended lab storage of specimens (Kalanetra et al. 2005). Consistent with this interpretation, in situ observations of Gulf of Mexico seep sediments show gradual spatial transitions of different populations within the same seep sediment area on a lateral scale of centimeters with orange filament-dominated mats in the center, followed by *Thiovulum*-like veils, and the white dots of *Thiomargarita* spp. on the periphery (Fig. 3b). Similar observations were made in the Guaymas Basin in the Gulf of California, where *Thiomargarita*-like cells occur mostly outside or on the periphery of the mat-covered hydrothermal hot spots (B. J. MacGregor, unpublished observations).

Sessile Filamentous "Candidatus **Marithrix"**

During a survey of the northern Gulf seep sites with R/V *Atlantis* in November 2010 (cruise AT18-02), push cores collected by HOV *Alvin* recovered fluffy tufts of unpigmented, large filaments, ca. 110 μm in diameter, that were growing in extensive patches on the sediment surface near a small brine lake surrounded by brine flows and small mud volcanoes (named "Dead Crab Lake", GC246, Fig. 4). The filaments occurred on the seawater-exposed sediment surface, and did not penetrate the sediment to any extent. Within the sediments, high sulfide and ammonia concentrations were measured at a few cm depth (ca. 2 and 4 mM, respectively) that rapidly decreased towards the surface. The sediments were rich in DOC (ca. 0.5–0.6 mM), depleted in nitrate (ca. 1 μM) and had a porewater pH between 7.5 and 8.2 (Salman-Carvalho et al. 2016). By 16S rRNA gene analysis, the filaments were members of a monophyletic branch within the *Beggiatoaceae* that was recently classified as "*Candidatus* Marithrix" (Salman-Carvalho et al. 2016; Salman et al. 2011). The "*Ca.* Marithrix" lineage contains large, filamentous vacuolated filamentous sulfur oxidizers, but in contrast to other large vacuolated

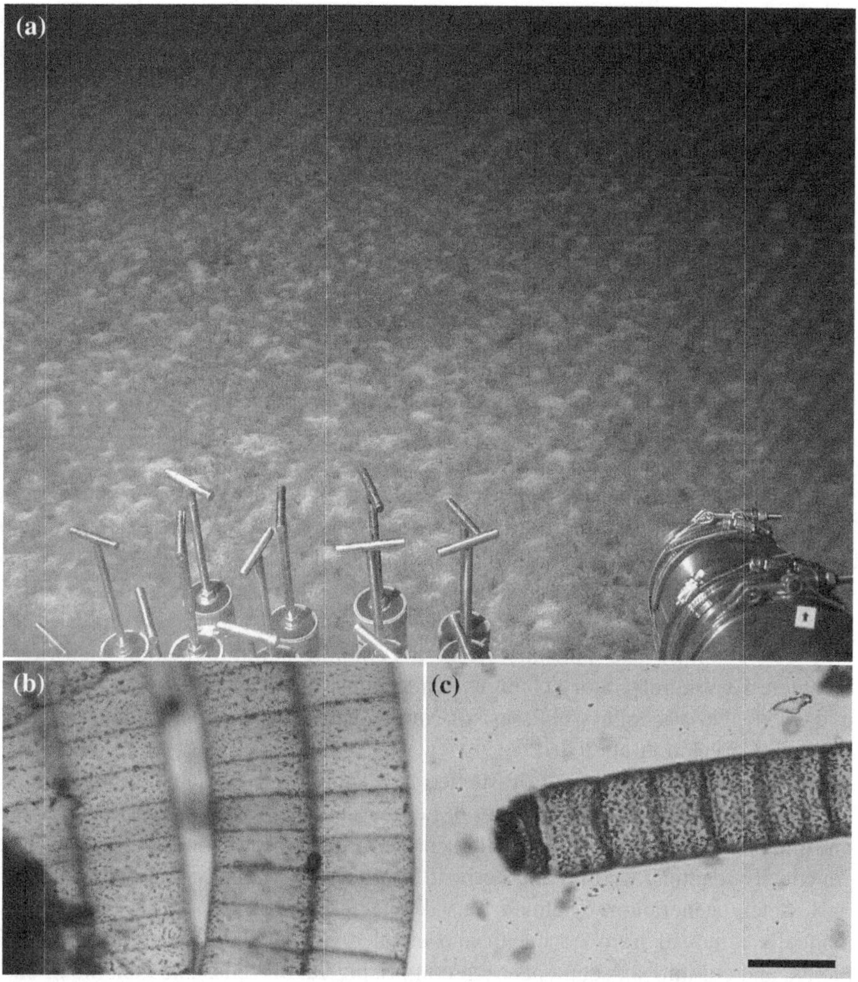

Fig. 4 "*Candidatus* Marithrix" in Gulf of Mexico sediments. **a** "*Ca*. Marithrix" sampling area at GC246, photographed before sampling during *Alvin* dive 4652, Nov. 23, 2010. The filaments grow on a light-colored sediment surface with occasional dark-grey patches, indicating strongly reducing conditions just underneath the surface. This site had disappeared when GC246 was revisited in April 2014, and patterns of reducing brine flows and seep areas had changed. **b** Microscopic image of several "*Ca*. Marithrix" filaments showing near-transparent, "hollow" cells with small (presumably sulfur) granules in the cytoplasm. Image taken by A. Teske, UNC. **c** Microscopic image of an isolated filament whose tip appears differentiated, reminiscent of a holdfast structure observed in *Thiothrix*. Images a and c from Salman-Carvalho et al. 2016. Scale bar = 20 μm

Beggiatoaceae, filaments in this phylogenetic group are frequently observed to attach to seawater-exposed surfaces (Kalanetra et al. 2004; Kalanetra and Nelson 2010; Salman-Carvalho et al. 2016). Filaments also attach to each other, forming patches, or rosettes that float in the bottom water (Kalanetra et al. 2004). The attachment is possibly realized by a holdfast structure formed at one end of the filament, analogous to differentiations known from *Thiothrix* organisms (Kalanetra et al. 2004; Kalanetra and Nelson 2010; Salman-Carvalho et al. 2016; Williams et al. 1987). "*Ca.* Marithrix" from locations other than the Gulf of Mexico were tested for the accumulation of intracellular nitrate, however, in contrast to the other vacuolated *Beggiatoaceae* this could not yet be demonstrated, despite the presence of the vacuolar structure inside the filaments (Kalanetra et al. 2004; Kalanetra and Nelson 2010). It remains unclear whether the filaments may only utilize nitrate directly from their surrounding medium without accumulating it, or, whether the filaments temporarily did not store nitrate at the time of sampling. Furthermore, it was experimentally shown that these organisms are capable of carbon fixation, indicating that they can live autotrophically (Kalanetra et al. 2004; Kalanetra and Nelson 2010). Genomic information from a Gulf of Mexico "*Ca.* Marithrix" filament indicates that they may utilize nitrate for facultative anaerobic respiration, and may be facultative autotrophic sulfide oxidizers. The genome encodes RuBisCo Form I, high-affinity carbon uptake proteins, the TCA cycle, glycolysis, and glycogen synthase and phosphorylase, a complete sulfur oxidation pathway, a complete aerobic respiratory chain with a terminal cytochrome c oxidase, however it lacked the vacuolar ATPases of other large vacuolated sulfur oxidizers (Salman-Carvalho et al. 2016).

In the same study, the Gulf of Mexico "*Ca.* Marithrix" filaments were also used to address a more conceptual question, i.e. the genomic complexity of cells within one filament. The cells within a filamentous organism are considered clonal, i.e. genetically identical, as they are derived by division from ancestral cells and span only a few generations (Salman-Carvalho et al. 2016). A single filament was manually removed from its sheath and cleaned of potentially contaminating epibionts, and cut into six segments each comprising ca. 30 cells. The genome of each segment was then separately amplified and sequenced. After alignment of the genome sequence reads the genetic content across the entire filament was largely identical. Comparing, however, the raw sequence reads within and across the segments, specific genome locations became apparent where conservation was reduced. These positions of high heterogeneity do not only reflect the potential differences across the 30 cells of a segment. As "*Ca.* Marithrix" and other large sulfur bacteria are highly polyploid they represent loci of potential microheterogeneity among genome copies within a single cell. Therefore, these positions might function as hotspots for evolutionary change and diversification of genomes belonging to newly separated daughter cells, and of genomes within in the same polyploid cell (Salman-Carvalho et al. 2016).

New Findings of *Beggiatoaceae* Types in the Gulf of Mexico

Discovery of Thioploca-like Organisms

One major morphotype known among the large, sulfur-oxidizing bacteria has not yet been described to occur around seep sites, i.e. filamentous, bundle-forming types, or, members of the genus *Thioploca*, or, the candidate group "Marithioploca". These *Beggiatoaceae* grow as bundles of multiple filaments with gliding motility within a mucilaginous sheath embedded several centimeters vertically in the sediment (Teske and Salman 2014). The lineage of large, vacuolated marine representatives of "*Ca.* Marithioploca", has been extensively studied in sediments of coastal upwelling areas on the Chilean continental shelf (e.g., Fossing et al. 1995; Gallardo 1977; Høgslund et al. 2009; Jørgensen and Gallardo 1999; Otte et al. 1999). Bundle-forming freshwater and brackish *Thioploca* are thinner and either do not feature internal vacuoles and nitrate storing capabilities, or can store nitrate at a much lesser extent than their large relatives (Høgslund et al. 2010; Kojima and Fukui 2003; Maier and Preissner 1979). Within the sheath, filaments shuttle between the oxygenated sediment surface where nitrate is taken up into the vacuoles, and the deeper sediment layers where sulfide is continuously oxidized and rapidly consumed, in spite of high sulfate reduction rates (Hüttel et al. 1996).

Here we report previously unpublished results from sediments collected by *Alvin* push coring with 16" *Alvin* cores at two locations: in the Megaplume area of Green Canyon 600 (27°N 22.174; 90°W 34.8289) at ca. 1225 m depth (Alvin Dive 4691, April 5, 2014), and near Dead Crab Lake at Green Canyon 246 (27°N 42.117; 90° W 38.893) at ca. 830 m depth (Alvin Dive 4694, April 10, 2014). These sediment cores harbored sheathed, filamentous bacteria in olive-brown, non-sulfidic surficial sediments down to 2–3 cm depth, but not in the reduced, dark-grey and sulfidic sediment layers below. Macroscopically, the bundles featured a major difference to known bundled sulfide-oxidizers of the genera *Thioploca* and "*Candidatus* Marithioploca": the sheaths of the bundles were strongly entangled and branched multiple times (Fig. 5a, b). Individual sulfur-filled filaments were ca. 18–20 μm in diameter and could be occasionally observed outside the sheath (Fig. 5c). Most sheaths were filled with numerous filaments (Fig. 5d). In order to phylogenetically identify the organism, fragments of filaments from different depths were pushed through 0.2% agarose to eliminate sheath material and other bacteria, and were frozen in sterile water for 16S rRNA gene sequencing when back on land; however, despite multiple attempts no 16S rRNA genes could be PCR-amplified. Occasionally, individual filaments were observed emerging from the bundle (Fig. 5c), indicating that they were motile and could move in and out of the sheath; however, the filaments showed no gliding motion at the time of shipboard observation. The large vacuolated "*Candidatus* Marithioploca" are known to be highly oxygen sensitive, and require anaerobic handling and incubation conditions (Otte et al. 1999). The absence of any filament motility even during the initial stages of

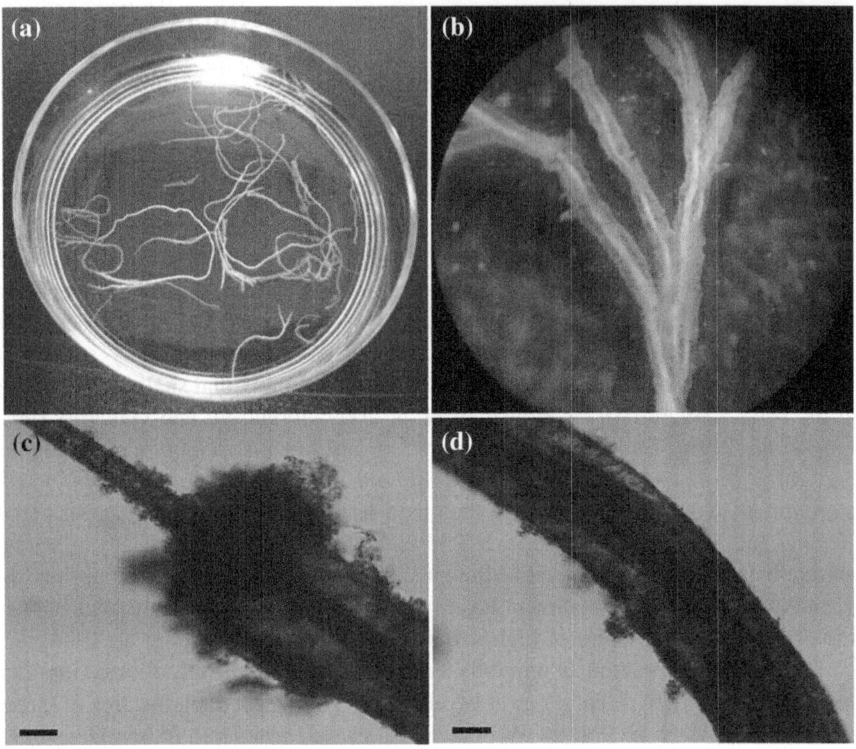

Fig. 5 Branched-bundled filamentous sulfur bacteria found in GC246 (core 4691-7). The sheaths (**a**) were macroscopically visible and varied in color from white, over yellow, to orange. The bundles were often tightly entangled and also branched (**b**). The filaments were occasionally emerging from the sheath (**c**), but were mainly observed completely retracted within the sheath (**d**). Images taken by V. Carvalho. Scale bars = 20 μm

shipboard examinations of the Gulf of Mexico *Thioploca*-like organisms indicates that the samples were already inactive at the time of sampling, or that damage occurred during transport to the surface.

Obviously, sampling at the right place and the right time is essential to track down active populations of these *Thioploca*-like populations, to study their phylogenetic affiliation, and their ecophysiological role. An interesting aspect of this discovery is that the branched *Thioploca*-like filaments were only found in sediments outside of microbial mats or seep areas, where the olive-brown seafloor showed no indication of white or pigmented mats, *Thiomargarita* cells, or dark-grey sulfides. During dive 4691, the branched filament bundles were found in olive-brown control cores 2 m distant from a microbial mat area with white, fine-grained sulfur "dusting", but not on the edge or within this sulfidic area, where fully reducing black-grey sediments dominated.

The filaments were found again during emergency sampling at Dead Crab Lake (GC246) on dive 4694 when *Alvin* was suddenly engulfed by a cloud of finely dispersed sediment, moving in from the direction of Dead Crab Lake, which at that point had evolved from the quiescent brine lake observed in 2010 (Stevens et al. 2015) into an intermittently erupting briny mud volcano. The decreased visibility forced *Alvin* to settle on the seafloor and take cores of olive-brown oxidized-looking sediments on the spot. These circumstances indicate that sediment sampling with a focus on microbial mats, sulfur deposits or conspicuous seep areas with fully reduced sediments could easily miss these *Thioploca*-like organisms. We hypothesize that sediments on the margins of seep areas, with significantly reduced seepage and inconspicuous oxidized surfaces overlying reduced sediments with residual low-activity seepage, provide their natural habitat (Fig. 6).

The identification of these bacteria as *Thioploca* spp. matches their geochemical habitat characteristics. Microelectrode profiling of the sediments with *Thioploca*-like filament bundles at Dead Crab Lake (cores 4694-8, -9, -10, and -11) showed consistently oxygen depletion at the sediment surface, the absence of sulfide throughout the upper 1.5 cm, and the gradual appearance and accumulation of sulfide around 2 cm depth (Fig. 7). A relative pH maximum near 8.5 at 2 to 2.5 cm depth, where oxygen was not available and ammonia concentrations are increasing, would be consistent with nitrate-reducing sulfide or sulfur oxidation producing nitrogen or ammonia:

$$4HS^- + NO_3^- + 6H^+ \rightarrow 4S^0 + NH_4^+ + 3H_2O$$
$$HS^- + NO_3^- + H^+ + H_2O \rightarrow SO_4^{2-} + NH_4^+$$
$$5HS^- + 2NO_3^- + 7H^+ \rightarrow 5S^0 + N_2 + 6H_2O$$
$$5HS^- + 8NO_3^- + 3H^+ \rightarrow 5SO_4^{2-} + 4\ N_2 + 4H_2O$$

The negative control core, 4691-15, from dark-grey, reducing sediments nearby, lacks this suboxic zone; here, sulfide forms a concentration peak at ca. 1 cm depth, and penetrates the surface sediment towards the sediment/water interface, where it overlaps with oxygen within a millimeter-thin gradient. These sediments, although sampled at a distance of only 2 m, lacked *Thioploca*-like filaments entirely.

The suboxic sediment layer coincides with the occurrence of *Thioploca*-like filaments in the upper 2 cm; when this suboxic sediment layer is missing, *Thioploca*-like filaments are not found. Thus, the evidence from sulfide, oxygen and pH profiles and filament bundle occurrence patterns suggests that the depth extent of the *Thioploca*-like filament bundles defines the sediment layer where sulfide is being anaerobically oxidized, possibly by nitrate reduction. High-resolution profiling of nitrate and ammonia concentrations would provide further evidence to test this working hypothesis.

Using the generally high ammonia concentrations of briny seep fluids in the Gulf of Mexico (Joye et al. 2010) as a proxy for seepage intensity, the *Thioploca*-like filament bundles were consistently found in moderately reducing sediments on the

(a)
4694-8
4694-9
4694-10
4694-11
4694-15
(b)
(c)

◂**Fig. 6** The sampling site at Dead Crab Lake containing *Thioploca*-like organisms. **a** Surficially oxidized, olive-colored sediments were cored as shown in this *Alvin* screengrab image (cores 8, 9, 10, 11, 13, taken in close proximity to each other). The sediment surface color transitions into dark-grey towards the *Alvin* sampling basket and the lower margin of the photo. *Alvin* dive 4694, April 10, 2014. **B** Following recovery and cleaning in the cold room, the cores showed a consistent olive-brown suboxic layer, ca. 2–3 cm thick (in stippled box) that transitioned sharply into dark-grey sulfidic sediment below. Bundled filamentous sulfur bacteria were found in the suboxic sediment layers of these cores. Core 15 **c** was taken ca. 2 m apart from the olive-green sediment site; it was fully reduced towards the surface and served as control core, containing no *Thioploca*-like filaments. Photos by *Alvin* crew, WHOI (**a**), and by A. Teske and B. J. MacGregor, UNC (**b** and **c**)

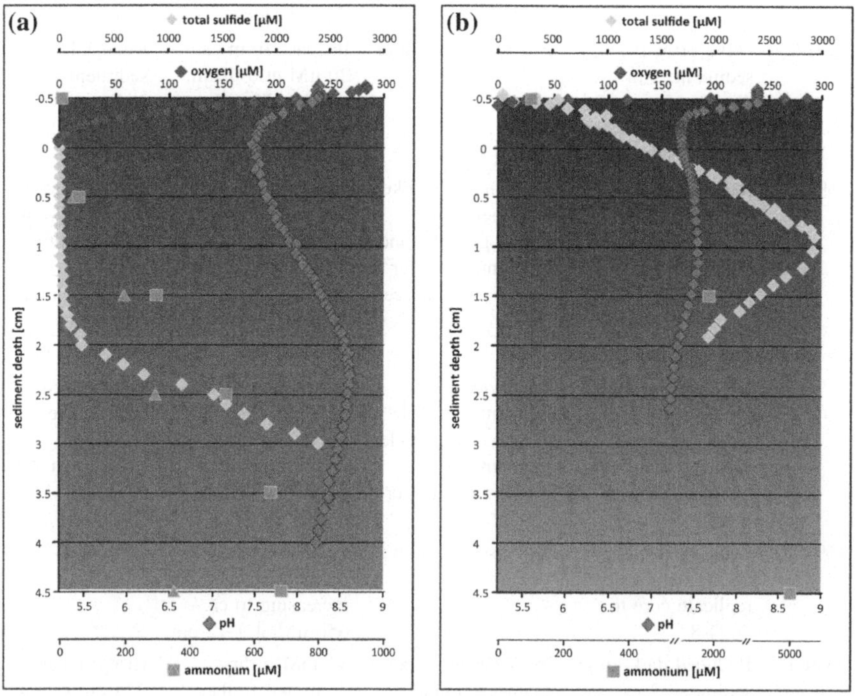

Fig. 7 Microprofiles in sediment cores containing branched-bundled filamentous *Thioploca*-like organisms (**a**), and fully reduced control site (**b**). **a** Oxygen and total sulfide concentrations, and pH values are averages across cores 4694-8, -9, -10, and -11, located adjacent to each other in olive-green sediment. The ammonia values are shown for core 10 (triangles) and 8 (squares; after 8 days of incubation with 1 mM NO_3^- added to the overlying water). **b** Control site with fully reduced, dark-grey sediment, where no *Thioploca*-like organisms could be found

margins of fully reduced seep areas. Porewater ammonia concentrations of incompletely reduced cores harboring *Thioploca*-like filaments in the surface sediment layers remained near 0.5–1 mM or less, and salinities were only moderately elevated compared to seawater, up to 47 permille total salinity (Table 1). For comparison, ammonia porewater concentrations of fully reduced, briny cores reached values of multiple mM (Joye et al. 2010, and Table 1).

Table 1 Summary of environmental conditions and the presence of *Thioploca*-like branched-bundled filamentous and other large sulfur bacteria. Specific observations and experimental variations for individual samples are explained in footnotes with asterisks

Dive and core no.	Sediment color, oxidation state, and mat cover	Observations of sulfur-oxidizing bacteria	Ammonia concentrations	Other core characteristics
Green Canyon 600, Megaplume area				
4691-3	Olive-brown sediment, 2 m distance from mat area	*Thioploca*-like filaments 0–1, 1–2 and 2–3 cm, few at 3–4 and 4–5 cm	Increase from ca. 30 μM at 0–2 cm, to 300 μM at 4–6 cm and >500 μM around 16 cm	Reduced sediment with carbonate chunks below 5 cm
4691-7[a]	Olive brown sediment, 2 m distance from mat area; replicate core to 4691-3	No observation data	Increase from ca. 20 μM at 0–1 cm, to 270 μM at 5–8 cm and 180 μM around 16 cm	Reduced sediment with carbonate chunks below 5 cm
4691-12	Reduced dark sediment, with white sulfur "mat", or sulfur precipitates	No *Thioploca*- like organisms, but mixed globular and filamentous sulfur bacteria on surface	Highest values of 210 μM in first cm, then stable at 40–50 μM after third cm	Fully reduced over top 7 cm; then transition to lighter-colored sediment
Green Canyon 246, Dead Crab Lake				
4694-8[b]	Olive-brown upper 2 cm, then black-grey and sulfidic; small gas bubbles below 8–9 cm	*Thioploca*-like filaments[c] in 0–1 and 1–2 cm; single filaments in 2–3 cm, none deeper	Increase from <10 μM at surface to >600 μM at 3 cm, increasing to >1 mM at 12 cm depth	Porewater a little above seawater salinity, ca. 40–47 ppt
4694-10	Olive-brown upper 2 cm, then reduced; replicate core to 4694-8	No observation data	Ca. 300 μM at 3 cm, then increasing to ca. 650 μM at 9–12 cm	
4694-16	Reduced dark-grey sediment, 2 m distance from olive-brown area	No *Thioploca*-like filaments found	>1 mM in first 3 cm and further increase to >5 mM downcore	Briny; reaches 225 ppt salinity at 12 cm depth

[a]After 5 days of cold room incubation and supernatant bubbled with air; this treatment sharpened the transition from olive-brown to dark-grey sediment at ca. 5 cm depth

[b]After 6 days of cold room incubation with 1 mM NO_3^- added to supernatant; this treatment sharpened the transition from olive-brown to dark-grey sediment at ca. 5 cm depth

[c]Filaments looked healthier, richer in white sulfur globules than those in the GC600 samples, possibly due to 6 days prior incubation with 1 mM NO_3^- addition before core dissection and filament sampling

Macroscopic Aggregates of Curled Filamentous Sulfide Oxidizers

Sediment cores that did not contain *Thioploca*-like filaments harbored other morphotypes of sulfur-oxidizing bacteria that became conspicuous after a few days of ex situ incubation. Jar incubations were performed with the surface sediments (top 1–2 cm) of cores that appeared promising for enriching sulfur bacteria; such cores contained either a few filaments or a white powdered surface when arriving on board, or their surface had been disturbed by outgassing. During *Alvin* dive 4691, cores 4 and 10 were taken within the sulfur-dusted mat areas at GC600; these sediments developed populations of morphologically distinct filamentous sulfur bacteria within a few days of incubation. While core 10 sediments developed different diameter classes of free-living, gliding filamentous sulfur oxidizers, the filaments enriched from core 4 formed tight, macroscopically visible aggregates (up to 1 cm in diameter) that appeared slightly pink (Fig. 8a, b). When the tightly wound, sticky filaments were carefully pulled apart, they were curled like a corkscrew (Fig. 8c), a previously undocumented morphology for large filamentous sulfur-oxidizing bacteria. Under the microscope, the filaments consisted of very narrow cells lined up like a stack of coins, and contained sulfur inclusions. Cells contained an unstained center upon FITC staining (Fig. 8d) and appeared hollow with brightfield microscopy (Fig. 8e), indicating internal vacuolation.

Bubbling of the overlying water for a few hours or even days resulted in an expansion of the aggregate, as the filaments protruded from the aggregate into the water. Also, additional small aggregates formed on the sediment surface. The filaments were non-motile when observed under the microscope, however, their extrusion from the main mass of the aggregate implies an active motility. Nitrate addition (100 mM) and simultaneous bubbling for 3 days had no visible effect on the filaments.

One aggregate was selected for microprofiling (Fig. 9). It had a diameter of ca. 2 mm and had not been previously manipulated, residing in the spot where it had formed. The overlying water was gently bubbled with air for several days before the measurement took place. The profiles show oxygen uptake and consumption around the aggregate, leaving the interior aggregate anoxic. Concentrations of sulfide were highly fluctuating when the sensor penetrated the aggregate, and thereafter also within the underlying sediment. These results indicate the buildup of sulfide towards the sediment surface and its penetration into the overlying aggregate, where the microorganisms that constituted the aggregate decreased the sulfide concentrations, most likely as a result of microbial sulfide oxidation. The uneven sulfide concentrations within the aggregate indicate the formation of micro-niches; the non-steady-state situation of sulfide below the aggregate is indicative of a dynamic movement of the filaments and local consumption of sulfide by the

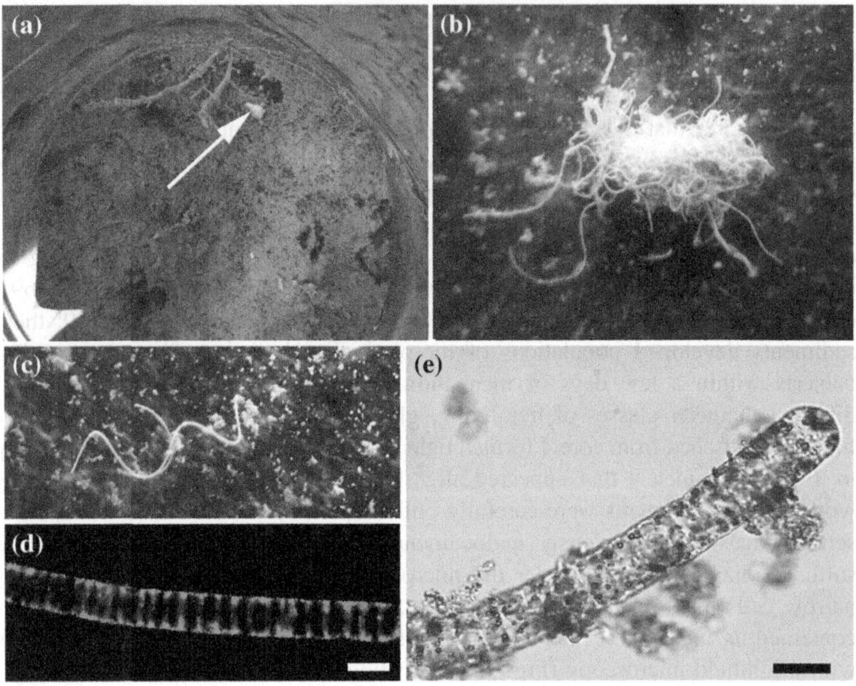

Fig. 8 A bird's eye view into the jar (**a**) shows the largest aggregate (arrow), which was recovered (**b**) and carefully pulled apart. Single filaments within the aggregate were predominantly curled like a corkscrew (**c**). They were possibly vacuolated as they contained an unstained center upon FITC staining (**d**) that is commonly used to stain the cytoplasm. The cytoplasm contains internal sulfur globules (**e**). Photos and microphotographs by V. Carvalho. Scale bars = 20 μm

bacteria. The sulfide flux in the sediment below the aggregate was higher than at the control site. The pH profile revealed a pH decrease within the biomass of the aggregate, suggesting the oxidation of reduced sulfur species to sulfuric acid.

Enriching Thiomargarita-Like Organisms

Sediments from GC600 were also used for shipboard incubations. Jar incubations with surface sediment from *Alvin* dive cores 4693-9 and 4695-10 revealed a population of single, spherical sulfur bacteria after 1 week, resembling *Thiomargarita* species (Fig. 10). These cells had also been observed in freshly retrieved cores from dive 4688 at GC600 a few days before. They were typically round or oval-shaped,

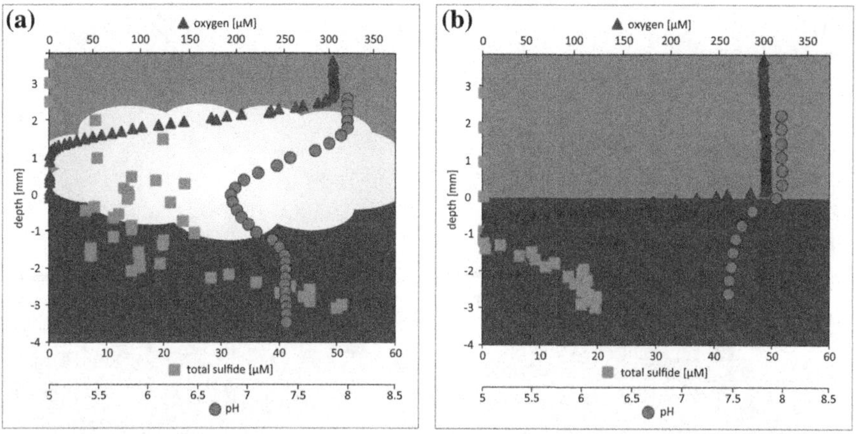

Fig. 9 Microelectrode profiles of sulfide, oxygen and pH through the aggregate consisting of curly filaments (**a**), and at a control site next to the aggregate (**b**). **a** Microprofiles through the aggregate showed that the aggregate consumed oxygen; at the same time sulfide reached the aggregate and was even detectable within it at highly fluctuating concentrations. The pH decreased within the aggregate, possibly reflecting acidification during sulfur oxidation. **b** Microprofiles of bare sediment next to the aggregate in the same jar show the conditions without the influence of aggregate-forming curly filaments. Oxygen reached the sediments and penetrated to about 1 mm depth, where it was fully depleted. Sulfide diffused upwards in the sediment and was depleted at 1 mm depth. The seawater pH of 8 decreased to ca. 7.5 within the sediment

ranged between 30 and 50 µm in diameter, and often had only relatively few sulfur inclusions (Fig. 10a, b). Others were very much filled with sulfur inclusions (Fig. 10c–e) and then had a very thick pronounced mucus sheath (Fig. 10d). Occasionally, cells with very large diameters (>300 µm; Fig. 10c) were observed. These unicellular large sulfur bacteria appeared sulfur-free in the center, implying the presence of a central aqueous vacuole, and they never showed mobility under the microscope. Often, cells in stages of binary fission were observed (Fig. 10e), and on three independent occasions structures that could resemble chain-formation were observed (Fig. 10f). This descriptive collection of *Thiomargarita*-like cells is slightly different from previously observed *Thiomargarita* in the Gulf of Mexico populations with most cells being smaller, and lacking the observations of blastula-like structure after multiple divisions in different spatial planes (Kalanetra et al. 2005). Jar-enriched populations possibly underwent mainly full separation of daughter cells after binary fission. Considering these differences, the jar-enriched population may not be in a state of famine, as suggested for the population studied by Kalanetra et al. (2005), or they may even represent another lineage of large, sulfur-oxidizing bacteria, for example members of the candidate groups "Thiopilula", or, "Thiophysa" (Salman et al. 2011; Fig. 1). No sequence-based identification of these large unicellular sulfur bacteria from the Gulf of Mexico is currently available.

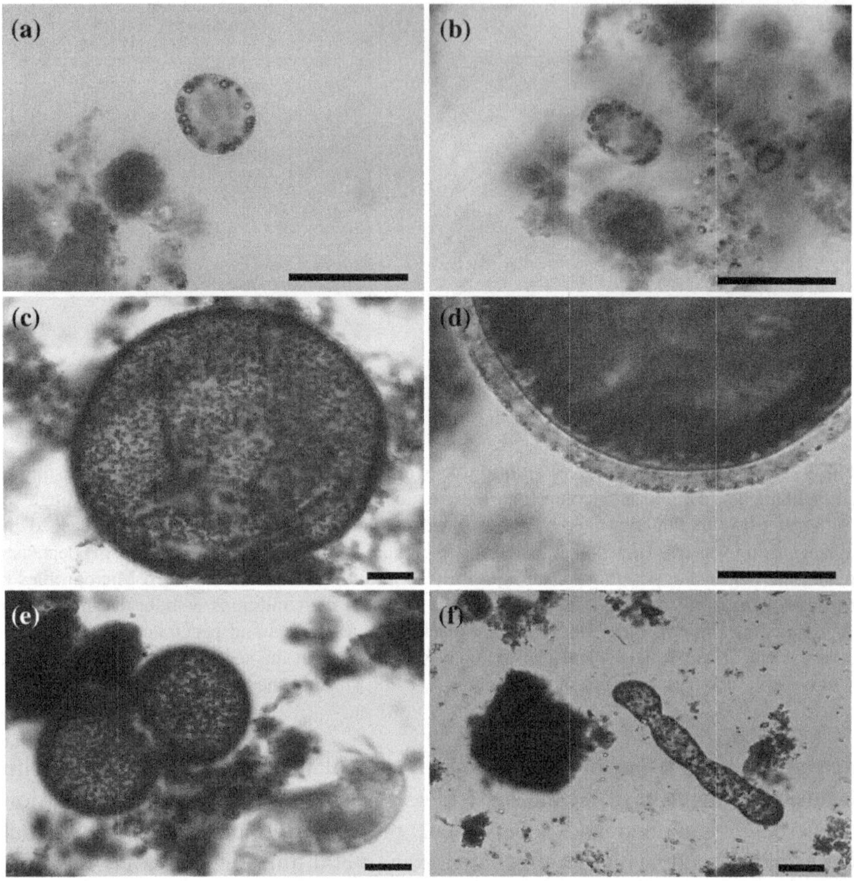

Fig. 10 Unicellular spherical large sulfur bacteria enriched in jars from core 4695-10. Most cells contained few sulfur inclusions and were round (**a**) or oval-shaped (**b**); some cells contained high amounts of sulfur (**c**–**e**). Cells were surrounded by a conspicuous outer sheath (**d**). Occasionally, very large cell sizes were observed (**c**); smaller cells showed binary fission stages (**e**). On three occasions, we observed a string of cells reminiscent of chain formation (**f**). Microphotographs by V. Carvalho. Scale bars = 50 μm

Conclusions

These preliminary findings indicate that seep sites in the Gulf of Mexico harbor numerous unidentified types of large sulfur-oxidizing bacteria. In many cases, these new types can be enriched to some extent from sediments lacking visible mats of conspicuous sulfur-oxidizing bacteria. Much remains to be learned about the dominant filamentous bacteria at Gulf of Mexico seeps, since they are clearly understudied compared to their cousins from the hot seep sites in Guaymas Basin. Surveying and identifying especially the less conspicuous types of large sulfur-oxidizing bacteria will extend the known biodiversity of this extensive microbial group, and help in understanding their specific microhabitats and ecological niches.

Acknowledgements The authors were supported by NSF Microbial Observatories-Microbial Interactions and Processes grant No. 0801742. We gratefully acknowledge the crew of R/V *Atlantis* and Submersible *Alvin* for exemplary support in the field during cruise AT18-02 in the Gulf of Mexico, and Chief Scientist Mandy Joye for steady leadership during a challenging cruise. Among the science crew, we thank in particular postdoctoral scientist Sairah Malkin for sharing microelectrode equipment and expertise.

References

Aharon P, Fu B (2000) Microbial sulfate reduction rates and sulfur and oxygen isotope fractionations at oil and gas seeps in deepwater Gulf of Mexico. Geochim Cosmochim Acta 64:233–246

Arvidson RS, Morse JW, Joye SB (2004) The sulfur biogeochemistry of chemosynthetic cold seep communities, Gulf of Mexico, USA. Mar Chem 87:97–119

Barry JP, Gary Greene H, Orange DL, Baxter CH, Robison BH, Kochevar RE, Nybakken JW et al (1996) Biologic and geologic characteristics of cold seeps in Monterey Bay, California. Deep Sea Res I 43:1739–1762

Fossing H, Gallardo VA, Jørgensen BB, Huttel M, Nielsen LP, Schulz H, Canfield DE et al (1995) Concentration and transport of nitrate by the mat-forming sulfur bacterium *Thioploca*. Nature 374:713–715

Gallardo VA (1977) Large benthic microbial communities in sulfide biota under Peru-Chile subsurface countercurrent. Nature 268:331–332

Guindon S, Dufayard J-F, Lefort V, Anisimova M, Hordijk W, Gascuel O (2010) New algorithms and methods to estimate maximum-likelihood phylogenies: assessing the performance of PhyML 3.0. Syst Biol 59:307–321

Hedrick DB, Pledger RD, White DC, Baross JA (1992) In situ microbial ecology of hydrothermal vent sediments. FEMS Microbiol Lett 101:1–10

Høgslund S, Revsbech NP, Kuenen JG, Jørgensen BB, Gallardo VA, Jvd Vossenberg, Nielsen JL et al (2009) Physiology and behaviour of marine *Thioploca*. ISME J 3:647–657

Høgslund S, Nielsen JL, Nielsen LP (2010) Distribution, ecology and molecular identification of *Thioploca* from Danish brackish water sediments. FEMS Microbiol Ecol 73:110–120

Hüttel M, Forster S, Kloser S, Fossing H (1996) Vertical migration in the sediment-dwelling sulfur bacteria *Thioploca* spp. in overcoming diffusion limitations. Appl Environ Microbiol 62:1863–1872

Jørgensen BB, Gallardo VA (1999) *Thioploca* spp.: filamentous sulfur bacteria with nitrate vacuoles. FEMS Microbiol Ecol 28:301–313

Joye SB, Boetius A, Orcutt BN, Montoya JP, Schulz HN, Erickson MJ, Lugo SK (2004) The anaerobic oxidation of methane and sulfate reduction in sediments from Gulf of Mexico cold seeps. Chem Geol 205:219–220

Joye SB, Bowles MW, Samarkin VA, Hunter KS, Niemann H (2010) Biogeochemical signatures and microbial activity of different cold-seep habitats along the Gulf of Mexico deep slope. Deep-Sea Res II 57:1990–2001

Kalanetra KM, Huston SL, Nelson DC (2004) Novel, attached, sulfur-oxidizing bacteria at shallow hydrothermal vents possess vacuoles not involved in respiratory nitrate accumulation. Appl Environ Microbiol 70:7487–7496

Kalanetra KM, Joye SB, Sunseri NR, Nelson DC (2005) Novel vacuolate sulfur bacteria from the Gulf of Mexico reproduce by reductive division in three dimensions. Environ Microbiol 7:1451–1460

Kalanetra KM, Nelson DC (2010) Vacuolate-attached filaments: highly productive *Ridgeia piscesae* epibionts at the Juan de Fuca hydrothermal vents. Mar Biol 157:791–800

Kalyaanamoorthy S, Minh BQ, Wong TKF, von Haeseler A, Jermiin LS (2017) ModelFinder: fast model selection for accurate phylogenetic estimates. Nat Methods 14:587

Kojima H, Fukui M (2003) Phylogenetic analysis of *Beggiatoa* spp. from organic rich sediment of Tokyo Bay, Japan. Water Res 37:3216–3223

Larkin JM, Henk MC (1996) Filamentous sulfide-oxidizing bacteria at hydrocarbon seeps of the Gulf of Mexico. Microsc Res Tech 33:23–31

Lloyd KG, Albert DB, Biddle JF, Chanton JP, Pizarro O, Teske A (2010) Spatial structure and activity of sedimentary microbial communities underlying a *Beggiatoa* spp. mat in a Gulf of Mexico hydrocarbon seep. PLoS ONE 5:e8738

Ludwig W, Strunk O, Westram R, Richter L, Meier H, Yadhukumar Buchner A et al (2004) ARB: a software environment for sequence data. Nucl Acids Res 32:1363–1371

MacDonald IR, Boland GS, Baker JS, Brooks JM, Kennicutt MC, Bidigare RR (1989) Gulf of Mexico hydrocarbon seep communities. Mar Biol 101:235–247

MacGregor BJ, Biddle JF, Siebert JR, Staunton E, Hegg EL, Matthysse AG, Teske A (2013) Why orange Guaymas Basin *Beggiatoa* spp. are orange: single-filament-genome-enabled identification of an abundant octaheme cytochrome with hydroxylamine oxidase, hydrazine oxidase, and nitrite reductase activities. Appl Environ Microbiol 79:1183–1190

Maier S, Preissner WC (1979) Occurrence of *Thioploca* in lake constance and Lower Saxony, Germany. Microb Ecol 5:117–119

McKay LJ, MacGregor BJ, Biddle JF, Albert DB, Mendlovitz HP, Hoer DR, Lipp JS et al (2012) Spatial heterogeneity and underlying geochemistry of phylogenetically diverse orange and white *Beggiatoa* mats in Guaymas Basin hydrothermal sediments. Deep-Sea Res I 67:21–31

Mills HJ, Martinez RJ, Story S, Sobecky PA (2004) Identification of members of the metabolically active microbial populations associated with *Beggiatoa* species mat communities from Gulf of Mexico cold-seep sediments. Appl Environ Microbiol 70:5447–5458

Minh BQ, Nguyen MAT, von Haeseler A (2013) Ultrafast approximation for phylogenetic bootstrap. Mol Biol Evol 30:1188–1195

Nguyen L-T, Schmidt HA, von Haeseler A, Minh BQ (2014) IQ-TREE: a fast and effective stochastic algorithm for estimating maximum-likelihood phylogenies. Mol Biol Evol 32:268–274

Nikolaus R, Ammerman JW, MacDonald IR (2003) Distinct pigmentation and trophic modes in *Beggiatoa* from hydrocarbon seeps in the Gulf of Mexico. Aquat Microb Ecol 32:85–93

Otte S, Kuenen JG, Nielsen LP, Paerl HW, Zopfi J, Schulz HN, Teske A, Strothmann B, Gallardo VA, Jørgensen BB (1999) Nitrogen, carbon, and sulfur metabolism in natural *Thioploca* samples. Appl Environ Microbiol 65:3148–3157

Quast C, Pruesse E, Yilmaz P, Gerken J, Schweer T, Yarza P, Peplies J et al (2013) The SILVA ribosomal RNA gene database project: improved data processing and web-based tools. Nucl Acids Res 41:D590–D596

Roberts HH, Aharon P, Carney R, Larkin J, Sassen R (1990) Sea floor responses to hydrocarbon seeps, Louisiana continental slope. Geo-Mar Lett 10:232–243

Roberts HH, Neurauter TW (1990) Direct observations of a large active mud vent on the Louisiana continental slope. AAPG Bulletin 74:1508

Salman-Carvalho V, Fadeev E, Joye SB, Teske A (2016) How clonal is clonal Genome plasticity across multicellular segments of a "*Candidatus* Marithrix sp." filament from sulfidic, briny seafloor sediments in the Gulf of Mexico. Front Microbiol 7:1173

Salman V, Amann R, Girnth A-C, Polerecky L, Bailey JV, Høgslund S, Jessen G et al (2011) A single-cell sequencing approach to the classification of large, vacuolated sulfur bacteria. Syst Appl Microbiol 34:243–259

Salman V, Bailey JV, Teske A (2013) Phylogenetic and morphologic complexity of giant sulphur bacteria. Antonie van Leeuw 104:169–186

Sassen R, Roberts HH, Aharon P, Larkin J, Chinn EW, Carney R (1993) Chemosynthetic bacterial mats at cold hydrocarbon seeps, Gulf of Mexico continental slope. Org Geochem 20:77–89

Schulz HN, Brinkhoff T, Ferdelman TG, Marine MH, Teske A, Jørgensen BB (1999) Dense populations of a giant sulfur bacterium in Namibian shelf sediments. Science 284:493–495

Schulz HN, Jørgensen BB (2001) Big bacteria. Annu Rev Microbiol 55:105–137

Sen Gupta BK, Platon E, Bernhard JM, Aharon P (1997) Foraminiferal colonization of hydrocarbon-seep bacterial mats and underlying sediment, Gulf of Mexico slope. J Foramin Res 27:292–300

Stevens EWN, Bailey JV, Flood BE, Jones DS, Gilhooly WP III, Joye SB, Teske A, Mason OU (2015) Barite encrustation of benthic sulfur-oxidizing bacteria at a marine cold seep. Geobiology 13:588–603

Teske A, Salman V (2014) The family *Beggiatoaceae*. In: Rosenberg E, Delong EF, Lory S, Stackebrandt E, Thompson FL (eds) The prokaryotes: gammaproteobacteria. Springer, Berlin-Heidelberg, pp 93–143

Williams TM, Unz RF, Doman JT (1987) Ultrastructure of *Thiothrix* spp. and "Type 021 N" bacteria. Appl Environ Microbiol 53:1560–1570

Wirsen CO, Jannasch HW, Molyneaux SJ (1992) Results of studies concerning microbiota. In: MacDonald IR, Schroeder W (eds). Chemosynthetic ecosystems study interim report, Appendix A. US Dept. Interior, minerals management service. Gulf of Mexico OCS Region, New Orleans, pp. A1–A14

Zhang CL, Huang Z, Cantu J, Pancost RD, Brigmon RL, Lyons TW, Sassen R (2005) Lipid biomarkers and carbon isotope signatures of a microbial *Beggiatoa* mat associated with gas hydrates in the Gulf of Mexico. Appl Environ Microbiol 71:2106–2112

Chapter 9
Growth Patterns of Giant Deep Sea *Beggiatoaceae* from a Guaymas Basin Vent Site

Dirk de Beer, Timothy Ferdelman, Barbara J. MacGregor, Andreas Teske and Charles A. Schutte

Abstract We studied the growth of giant filamentous sulfur oxidizing bacteria of the family *Beggiatoaceae* collected from a hydrothermal seep area in the Guaymas Basin. We measured the incorporation of ^{14}C-bicarbonate tracer into individual filaments using a microimager that allows quantitative determination of the distribution of radioisotopes with 20 μm resolution. Filaments incorporated label along their entire length; thus growth occurred uniformly throughout these whole filaments and not only at their tips. Uptake of ^{14}C-bicarbonate was strongly stimulated by reducing the pH from 8.2, the value near the sediment surface, to 7.05, as found within 1–2 mm below the surface; the presence of oxygen or sulfide had no effect. Thus, *Beggiatoaceae* strongly prefer assimilation of CO_2 over other DIC species. In consequence, migration of these motile filaments into deeper sediments, where sharply decreasing pH increases the availability of CO_2, will favor cell growth. Genomic evidence was found for periplasmic carbonic anhydrases, indicative of the carbon concentration mechanism.

D. de Beer (✉) · T. Ferdelman
Max-Planck-Institute Marine Microbiology, Bremen, Germany
e-mail: dbeer@mpi-bremen.de

T. Ferdelman
e-mail: tferdelm@mpi-bremen.de

B. J. MacGregor
University of Minnesota, Minneapolis, MN, USA
e-mail: macgregb@umn.edu

A. Teske
University of North Carolina at Chapel Hill, Chapel Hill, NC, USA
e-mail: teske@email.unc.edu

C. A. Schutte
Rowan University, Glassboro, NJ, USA
e-mail: schutte@rowan.edu

A. Teske and V. Carvalho (eds.), *Marine Hydrocarbon Seeps*,
Springer Oceanography, https://doi.org/10.1007/978-3-030-34827-4_9

Autotrophic CO_2 Fixation in *Beggiatoaceae*: Gaps in the Evidence

The family *Beggiatoaceae* is a diverse group of sulfur oxidizing bacteria that inhabit the upper sediments of reduced aquatic ecosystems (Teske and Nelson 2006). The large, gliding, vacuolated filamentous sulfur-oxidizers form an expanding collection of genus-level candidatus groups within the family *Beggiatoaceae* (Salman et al. 2011), and are commonly found in marine habitats, including "Parabeggiatoa", "Marithioploca", "Halobeggiatoa", and "Isobeggiatoa" (MacGregor 2016; Teske and Salman 2014). Many of these have until recently been wrongly referred to as *Beggiatoa* spp., although they are distinct from the fresh-water isolate *Beggiatoa alba*, which is not vacuolated (Mezzino et al. 1984). We leave the discussion on taxonomy to others and refer to the filaments presented here as 'giant *Beggiatoaceae*'.

Phylogenetic diversity accommodates metabolic heterogeneity within the *Beggiatoaceae*; filamentous large sulfur bacteria include autotrophic organisms that use RuBisCO for CO_2 fixation (Nelson and Jannasch 1983), strains that are capable of mixotrophic growth (Hagen and Nelson 1997), and isolates with heterotrophic metabolic capacity that possess the TCA cycle (Nelson and Castenholz 1981). Giant filamentous sulfur bacteria are abundant at hot and cold seeps, where they oxidize sulfide with oxygen or nitrate, and are likely important primary producers sustaining these ecosystems (Nelson et al. 1989). The diameter of the filaments can reach more than 100 μm, since each cell contains a large vacuole that stores nitrate at concentrations of up to several hundred mM (McHatton et al. 1996). This nitrate is used as a terminal electron acceptor in anoxic sediments, and where tested is estimated to be sufficient for cells to survive anoxia for several weeks (Preisler et al. 2007). As the filaments can deplete sulfide within the upper 2–3 cm of sediments, their motility and anaerobic metabolism carves out a special niche, increasing their competitiveness with single-celled sulfide oxidizers that cannot reach the sulfide pool.

Since cultures of marine *Beggiatoaceae*, including giant *Beggiatoaceae*, are currently not available, advances in physiological knowledge largely depend on enrichments obtained in physical separations (Nelson et al. 1989; Otte et al. 1999; Preisler et al. 2007; Sweerts et al. 1990) or in artificial sulfide-oxygen gradients (Nelson et al. 1986a, b; Nelson and Hagen 1995). These limitations do not allow typical microbiology studies on growth rates and kinetics, and only a few studies of growth patterns exist for filamentous *Beggiatoaceae*. One study was done using gradient cultures on a marine strain that is now regrettably lost; sophisticated mass balancing with microsensors and protein accumulations showed that these filaments oxidized sulfide with oxygen at a high rate and growth yield (Nelson et al. 1986a). Multiplication of filaments and cell division was studied in heterotrophically maintained cultures of *Beggiatoa alba*, with moderate filament diameters of 2–3 μm (Strohl and Larkin 1978). Cell divisions were observed randomly within the filaments, and sacrificial breakage of individual cells led to division of filaments into two pieces. No comparable data, however, are available on autotrophic growth patterns of vacuolated giant *Beggiatoaceae*, despite their importance as chemosynthetic primary producers in seep and vent habitats.

We sought to fill this gap by using microimaging of radioactivity to study ^{14}C incorporation in giant *Beggiatoaceae* with a diameter of 100 μm, to find out where the filaments grow, and to determine the effects of pH, sulfide and oxygen on autotrophic growth. We expected a homogenously distributed growth pattern along the filaments, a higher CO_2 fixation rate at lower pH due to shifts in the carbonate system towards higher CO_2 concentrations, and an increase in CO_2 fixation in the presence of sulfide. We also checked currently available genomes of Guaymas *Beggiatoaceae* for genes that encode periplasmic carbonic anhydrase (pCA), an essential enzyme in the carbon concentrating mechanism (CCM).

Experiments with Filamentous *Beggiatoaceae* from Guaymas Basin, Gulf of California

Samples of white *Beggiatoaceae* mats from the hydrothermal sediments of Guaymas Basin in the Gulf of California (McKay et al. 2012; Teske et al. 2016) were collected using push cores during cruise AT37-06 with R/V *Atlantis* and deep-sea submersible *Alvin*. Freshly recovered samples were transferred to the ship's cold room (4 °C). Filaments were gently pulled off the top of sediment cores using a pipette, and placed in 100 ml plastic beakers containing seawater. Some sediment adhered to the filaments that, when left overnight, ended up at the bottom of the beaker while the filaments reformed a mat on top of it. This transfer was repeated several times, to clean and enrich the filaments as much as possible before they were used in experiments (Schutte et al. 2018). One of these batches survived the transport to the laboratory in Bremen where the experiments were performed.

Filaments were incubated with radiotracer in 3 mL exetainers (Labco, UK) that contained 1 mL of seawater enriched with nitrate to a final concentration of 25 μM. Before adding tracer or filaments, the conditions were adjusted to one of four experimental conditions: oxic pH 8.2, oxic pH 7.05, anoxic pH 7.05, or sulfidic pH 7.05. Three replicate incubations were performed for each condition. Before injection into the exetainers, seawater was adjusted to pH 7.05 by flushing first with 10% CO_2 in N_2, followed by flushing with N_2 until the correct pH was reached. The sulfide was added to a final concentration of 30 μM from a stock that was adjusted to pH 7. A small tuft of around 20 *Beggiatoaceae* filaments was added and, for anoxic incubations, the head space was quickly replaced with N_2. The incubations were started by adding 100 kBq of ^{14}C-bicarbonate. The incubations were performed on melting ice (to approximate the in situ temperature of ~3 °C) and lasted approximately 8.5 h, after which 25 μL were taken to determine the specific activity (approximately 2.3×10^{12} CPM/mole, CPM being counts per minute). Then, the contents of each exetainer were filtered through gauze from a plankton net and rinsed 3 times with 5 mL seawater. Subsequently the filaments were distributed on polylysine slides (Thermo Fisher Scientific, Bremen, Germany) and quickly dried at 60 °C in a ventilated oven, on a metal support for efficient conduction.

Imaging was performed with a BetaImager (Biospacelab, Nesles la Vallée, France), and resulting images were analyzed using the ImageQuant software from the same company. The spiked filaments were selected and their areal CPM was determined. Areal carbon fixation rates in the labeled filaments were calculated using the specific activity of the bicarbonate and the CPM per area. Several samples were placed in a desiccator with a beaker of fuming HCl and imaged again to test for remaining unassimilated bicarbonate tracer, but this treatment did not reduce the signal further.

Microprofiles of O_2, pH, H_2S and ORP (oxidation reduction potential) were measured in situ as described previously (de Beer et al. 2006) using an autonomous microsensor profiler that was deployed and started by *Alvin*. The profiles were analyzed for seepage parameters as described (de Beer et al. 2006). In brief, total sulfide (S_{tot}) profiles were calculated from pH and H_2S profiles. The deeper part of the profiles (below the concentration maximum where no conversions occur) is only subjected to transport, where sulfide diffuses downward from a peak and is transported upwards by advection. As in steady state the local fluxes (J_x, where x is depth in the sediment) by diffusion equal the local fluxes by advection ($J_{Dx} = J_{ADVx}$), and $J_{Dx} = D \times dC_x/x$ and $J_{ADVx} = C_x \times v$, the upflow velocity is obtained, as $v = J_{Dx}/C_x$.

Genomic analyses were performed as described previously (Schutte et al. 2018). Protein domains were identified based on the Conserved Domain Database (Marchler-Bauer et al. 2017). Signal peptides were identified using the SignalP 4.1 Server (Petersen et al. 2011). Non-classical export signal sequence predictions were performed with the SecretomeP 2.0a Server (Bendtsen et al. 2005). For bacterial sequences, the SecP score should exceed 0.5 and there should additionally be no classic signal peptide. Transmembrane helixes were predicted using the TMHMM 2.0c Server (Krogh et al. 2001).

Inorganic Carbon Assimilation by Giant Guaymas *Beggiatoaceae*

Most filaments were labeled homogeneously along their length (Fig. 1). Those filaments that were only partially labeled showed no preferential pattern of label incorporation along the filament. Specifically, the filaments were not labeled predominantly at their end sections. Growth of large *Beggiatoaceae* from Guaymas Basin is therefore distributed evenly along individual filaments. However, growth rates varied substantially between filaments, and non-labeled filaments were observed in every preparation.

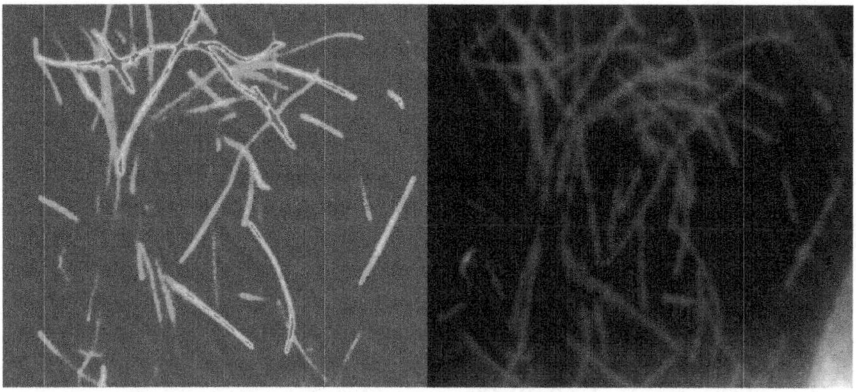

Fig. 1 An example of the data obtained from the BetaImager showing the distribution of radioactivity in *Beggiatoaceae* filaments after exposure to ^{14}C-bicarbonate tracer at pH 7.05 in the presence of 30 μM sulfide (left), and a light image of the preparation (right). Note that the light image is blurred because it was taken through the scintillation foil

Of the variables tested, only pH controlled the CO_2 fixation rate (Fig. 2). One unit decrease in pH led to approximately one order of magnitude faster CO_2 fixation. Moreover, the fraction of labeled cells doubled from 30 to 60%.

Sulfide addition did not increase growth, indicating that electron donor availability was not limiting. The sulfur granules stored in these filaments evidently provided an ample electron donor supply. The similar rates of label incorporation under oxic and anoxic conditions showed that the vacuolar nitrate reserve accepted electrons efficiently. Altogether, the filaments were sufficiently supplied with energy reserves in the form of nitrate and sulfur. The growth rate appears to be limited by dissolved inorganic carbon (DIC) uptake, which is thought to occur primarily by passive CO_2 diffusion across the membrane. The CO_2 concentration increases by a factor of ~14 upon acidification from pH 8.2 to 7.05, assuming the

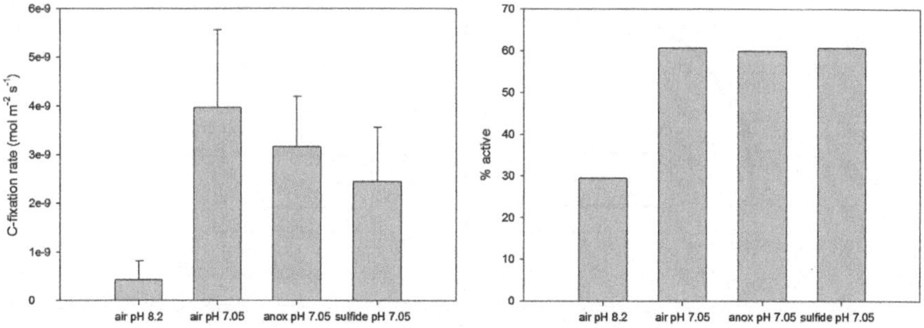

Fig. 2 Carbon fixation activity measured by ^{14}C-bicarbonate tracer. The fixation rates were expressed per area of flattened filaments (left). The percentage of active filaments (% active) were assessed by comparing the radiographic and light images (right)

total DIC concentration remains constant. Thus, the CO_2 concentration increase is of the same order of magnitude as the observed increase in the CO_2 fixation rate.

Microprofiles were measured in situ during a previous cruise (Teske et al. 2016) and showed a small but significant pH peak at the mat surface, bringing the pH 0.2 units above seawater levels (Fig. 3). Below this maximum the pH rapidly decreased to less than 7 within 2 mm. Oxygen concentrations were low and variable, such that nitrate stored in vacuoles is expected to be essential for the survival of filaments below the mat surface. The upflow velocity was calculated from the S_{tot} profile between 0.02 and 0.07 m depth, and was found to be approximately 2 m per year. As the upflowing porewater has high DIC concentrations averaging 20 mM (McKay et al. 2012, 2016) and low pH, it is highly enriched in CO_2 (3 mM at pH 7). Thus, within millimeters below the seafloor conditions become highly favorable for CO_2 uptake and CO_2 fixation by RuBisCO.

Despite high CO_2 concentrations in the *Beggiatoaceae* microhabitat, genomic evidence for the presence of a carbon concentrating mechanism (CCM) was found. The metagenome-derived genomes of white filaments (Beggiatoa. sp. bin 4484_5_27, and ex4572_84_Beggiatoa; Dombrowski et al. 2017) each possess genes for three different carbonic anhydrases (CA), enzymes that catalyze the transformation of CO_2 to HCO_3^- All three genes—two carbonic anhydrases and

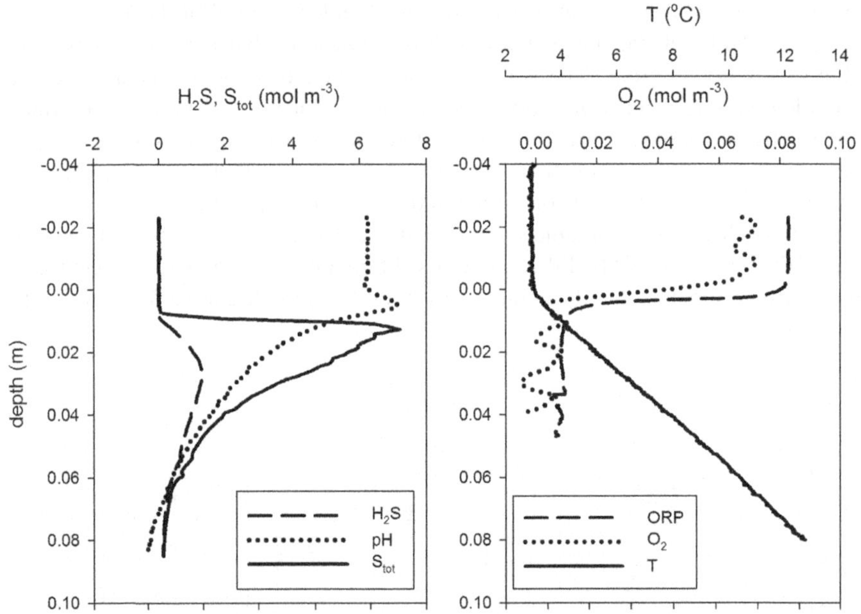

Fig. 3 Profiles of sulfide, oxygen, pH, and oxidation reduction potential (ORP) measured in situ on a white microbial mat in the Guaymas Basin. The total sulfide was obtained from the H_2S and pH profiles

one member of the carbonic anhydrase superfamily (Table 1)—belong to the α-type carbonic anhydrases that, at least in the well-studied hydrothermal vent sulfur-oxidizing bacterium *Thiomicrospira crunogena*, are localized in the periplasm (Dobrinski et al. 2010). The location of this pCA is important, as can be appreciated from the detailed mechanism of the CCM. A crucial step in the CCM is the active uptake of HCO_3^- from the periplasm into the cytoplasm, where it is converted to CO_2 via intracellular CA. Due to the CCM, CO_2 reaches high concentrations in the cytoplasm, which is needed to compensate for the low affinity of RuBisCO, the enzyme that converts CO_2 to organic carbon. The intracellularly produced CO_2 partially leaks out (Tchernov et al. 1997), reducing the efficiency of the CCM. The counterintuitive consequences of the CCM are elevated CO_2 and depleted HCO_3^- in the periplasm. The function of pCA is to convert CO_2 to HCO_3^-, the substrate for active DIC uptake. Indeed, insufficient HCO_3^- concentrations in the periplasm can limit DIC uptake if extracellular CA is absent or inactivated (Li et al. 2018). As active transport costs energy and CO_2 is the primary substrate for RuBisCO, high CO_2 concentration does stimulate carbon fixation even if a CCM is present (Wu et al. 2010). The CCM in *Beggiatoaceae* is probably only active at the sediment surface, where the conditions are close to those in bottom water and CO_2 concentrations are significantly lower than within the sediment.

It is generally accepted that gliding motility allows giant *Beggiatoaceae* to migrate between sulfide and nitrate pools, and it is speculated that motility helps to

Table 1 Carbonic anhydrases predicted for white Guaymas *Beggiatoaceae* filaments, from metagenome-derived genomes (Dombrowski et al. 2017; Schutte et al. 2018), based on full-length or partial amino acid sequences. Enzymes 1 and 2 were predicted as secreted proteins based on post-translational and localizational protein features (Bendtsen et al. 2005). The third enzyme differed from the others by having a predicted signal peptide and one transmembrane helix

Genome	Locus tag	Carbonic anhydrase type	Amino acid sequence length
Carbonic anhydrase Enzyme 1			
ex4572_84_Beggiatoa	Ga0123547_10073	Alpha	293
Beggiatoa sp. bin 4484_5_27	Ga0136995_106511	Alpha	293
Carbonic anhydrase Enzyme 2			
Beggiatoa. sp. bin 4484_5_27	Ga0136995_14805	Alpha	188
ex4572_84_Beggiatoa	Ga0123547_113615	Alpha (partial)	75
ex4572_84_Beggiatoa	Ga0123547_113616	Alpha (partial)	164
Carbonic anhydrase Superfamily member			
ex4572_84_Beggiatoa	Ga0123547_103312	Alpha with N-terminal NHL repeats	602
Beggiatoa sp. bin 4484_5_27	Ga0136995_10288	Alpha with N-terminal NHL repeats	602

escape predation (Van Gaever et al. 2006). Our results indicate that giant *Beggiatoaceae* may also use their gliding motility to access high levels of CO_2 in the underlying sediments to promote autotrophic carbon fixation and cellular growth.

Acknowledgements We are grateful for the excellent jobs done by the *Alvin* team and the R/V *Atlantis* crew during cruise AT37-06. Dirk de Beer's and Charles Schutte's cruise participation was financed by the Max-Planck-Society in Munich, Germany. The cruise was supported by NSF-Biological Oceanography (NSF-OCE 1357238).

References

Bendtsen JD, Kiemer L, Fausbøll A, Brunak S (2005) Non-classical protein secretion in bacteria. BMC Microbiol 5:58–70

de Beer D, Sauter E, Niemann H, Kaul N, Foucher JP, Witte U, Schlüter M, Boetius A (2006) In situ fluxes and zonation of microbial activity in surface sediments of the Håkon Mosby Mud Volcano. Limnol Oceanogr 51(3):1315–1331

Dobrinski KP, Boller AJ, Scott KM (2010) Expression and function of four carbonic anhydrase homologs in the deep-sea chemolithoautotrophc *Thiomicrospira crunogena*. Appl Environ Microbiol 76:3561–3567

Dombrowski N, Seitz KW, Teske AP, Baker BJ (2017) Genomic insights into potential interdependencies in microbial hydrocarbon and nutrient cycling in hydrothermal sediments. Microbiome 5:106

Hagen KD, Nelson DC (1997) Use of reduced sulfur compounds by *Beggiatoa* spp.: enzymology and physiology of marine and freshwater strains in homogeneous and gradient cultures. Appl Environ Microbiol 63:3957–3964

Krogh A, Larsson B, von Heijne G, Sonnhammer ELL (2001) Predicting transmembrane protein topology with a hidden Markov model: application to complete genomes. J Mol Biol 305:567–580

Li T, Sharp C, Ataeian M, Strous M, de Beer D (2018) Role of extracellular carbonic anhydrase in dissolved inorganic carbon uptake in alkaliphilic phototrophic biofilm. Front Microbiol 9:2490

MacGregor BJ (2016) Visualizing evolutionary relationships of multidomain proteins: an example from receiver (REC) domains of sensor histidine kinases in the *Candidatus* Maribeggiatoa str. Orange Guaymas Draft Genome. Front Microbiol 7:1780

Marchler-Bauer A, Bo Y, Han LY, He JE, Lanczycki CJ, Lu SN, Chitsaz F, Derbyshire MK, Geer RC, Gonzales NR, Gwadz M, Hurwitz DI, Lu F, Marchler GH, Song JS, Thanki N, Wang ZX, Yamashita RA, Zhang DC, Zheng CJ, Geer LY, Bryant SH (2017) CDD/SPARCLE: functional classification of proteins via subfamily domain architectures. Nucleic Acids Res 45(D1):D200–D203

McHatton SC, Barry JP, Jannasch HW, Nelson DC (1996) Nigh nitrate concentrations in vacuolate, autotrophic marine *Beggiatoa* spp. Appl Environ Microbiol 62:954–958

McKay LJ, MacGregor BJ, Biddle JF, Mendlovitz HP, Hoer D, Lipp JS, Lloyd KG, Teske AP (2012) Spatial heterogeneity and underlying geochemistry of phylogenetically diverse orange and white Beggiatoa mats in Guaymas Basin hydrothermal sediments. Deep-Sea Res Part 1 67:21–31

McKay LJ, Klokman VW, Mendlovitz HP, LaRowe DE, Hoer D, Albert DB, Amend JP, Teske A (2016) Thermal and geochemical influences on microbial biogeography in the hydrothermal sediments of Guaymas Basin, Gulf of California. Environ Microbiol Rep 8:150–161

Mezzino M, Strohl WR, Larkin JM (1984) Characterization of *Beggiatoa alba*. Arch Microbiol 137:139–144

Nelson DC, Castenholz RW (1981) Organic nutrition by *Beggiatoa* sp. J Bact 147:236–247

Nelson DC, Hagen KD (1995) Physiology and biochemistry of symbiotic and free-living chemoautotrophic sulfur bacteria. Am Zool 35:91–101
Nelson DC, Jannasch HW (1983) Chemoautotrophic growth of a marine *Beggiatoa* in sulfide-gradient cultures. Arch Microbiol 136:262–269
Nelson DC, Jørgensen BB, Revsbech NP (1986a) Growth pattern and yield of a chemoautotrophic *Beggiatoa* sp. in oxygen-sulfide microgradients. Appl Environ Microbiol 52:225–233
Nelson DC, Revsbech NP, Jørgensen BB (1986b) Microoxic-anoxic niche of *Beggiatoa* spp.: microelectrode survey of marine and freshwater strains. Appl Environ Microbiol 52:161–168
Nelson DC, Wirsen CO, Jannasch HW (1989) Characterization of large, autotrophic *Beggiatoa* spp. abundant at hydrothermal vents of the Guaymas Basin. Appl Environ Microbiol 55:2909–2917
Otte S, Kuenen JG, Nielsen LP, Paerl HW, Zopfi J, Schulz HN, Teske A, Strotmann B, Gallardo VA, Jørgensen BB (1999) Nitrogen, carbon, and sulfur metabolism in natural *Thioploca* samples. Appl Environ Microbiol 65:3148–3157
Petersen TN, Brunak S, von Heijne G, Nielsen H (2011) SignalP 4.0: discriminating signal peptides from transmembrane regions. Nat Methods 8(10):785–786
Preisler A, de Beer D, Lichtschlag A, Lavik G, Boetius A, Jørgensen BB (2007) Biological and chemical sulfide oxidation in a *Beggiatoa* inhabited marine sediment. ISME J 1:341–353
Salman V, Amann R, Girnth A-C, Polerecky L, Bailey J, Høgslund S, Jessen G, Pantoja S, Schulz-Vogt HN (2011) A single-cell sequencing approach to the classification of large, vacuolated sulfur bacteria. Syst Appl Microbiol 34:243–259
Schutte C, Teske A, MacGregor B, Salman-Carvalho V, Lavik G, Hach P, deBeer D (2018) Filamentous giant *Beggiatoaceae* from Guaymas Basin are capable of both denitrification and dissimilatory nitrate reduction to ammonium (DNRA). Appl Environ Microbiol 84:e02860-17
Strohl WR, Larkin JM (1978) Cell division and trichome breakage in *Beggiatoa*. Curr Microbiol 1:151–155
Sweerts J-PRA, de Beer D, Nielsen LP, Verdouw H, van den Heuvel JC, Cohen Y, Cappenberg TE (1990) Denitrification by sulphur oxidizing *Beggiatoa* spp. mats on freshwater sediments. Nature 344:762–763
Tchernov D, Hassidim M, Luz B, Sukenik A, Reinhold L, Kaplan A (1997) Sustained net CO_2 evolution during photosynthesis by marine microorganism. Curr Biol 7:723–728
Teske A, Nelson DC (2006) The genera *Beggiatoa* and *Thioploca*. In: Dworkin M, Falkow S, Rosenberg E, Schleifer KH, Stackebrandt E (eds) The Prokaryotes, vol 6. Springer, New York, pp 784–810
Teske A, Salman V (2014) The family *Beggiatoaceae*. In: Rosenberg E, DeLong EF, Thompson F, Lory S, Stackebrandt E (eds) The Prokaryotes—Gammaproteobacteria. The Prokaryotes, 4th edn. Springer, Berlin, pp 93–134
Teske A, de Beer D, McKay LJ, Tivey MK, Biddle JF, Hoer D, Lloyd KG, Lever MA, Røy H, Albert DB, Mendlovitz HP, MacGregor BJ (2016) The Guaymas Basin hiking guide to hydrothermal mounds, chimneys, and microbial mats: complex seafloor expressions of subsurface hydrothermal circulation. Front Microbiol 7:1–23
Van Gaever S, Moodley L, de Beer D, Vanreusel A (2006) Meiobenthos at the Arctic Håkon Mosby Mud Volcano, with a parental-caring nematode thriving in sulphide-rich sediments. Mar Ecol Prog Ser 321:143–155
Wu Y, Gao K, Riebesell U (2010) CO_2-induced seawater acidification affects physiological performance of the marine diatom *Phaeodactylum tricornutum*. Biogeosciences 7:2915–2923

Chapter 10
Uncovering Microbial Hydrocarbon Degradation Processes: The Promise of Stable Isotope Probing

Tony Gutierrez and Sara Kleindienst

Abstract Traditional microbiological methods for the identification of microorganisms after they have been isolated in pure culture have revealed key players in the degradation of hydrocarbons. But have we identified them all? The conspicuous enrichment of an uncultured *Oceanospirillales* in a sub-surface hydrocarbon plume during the Deepwater Horizon oil spill is one of many examples highlighting that we are not there yet in this respect. Culture-dependent methods typically miss identifying 99% of microorganisms originating from environmental samples, and are on their own ineffective in resolving the diversity and function of natural microbial communities. Stable isotope probing (SIP) is a technique used to identify a target group of microorganisms which can actively metabolize a specific substrate in an environmental sample and, thus, under in situ-like conditions. The technique involves incubating an environmental sample with an isotopically-labeled (e.g., ^{13}C, or ^{15}N) substrate and allowing the label to become incorporated into the biomass (e.g. DNA, RNA, protein, PLFAs) of those microorganisms capable of metabolizing the substrate. The labeled biomolecules are then isolated and analyzed to identify the organisms that actively incorporated the isotope label. SIP based on DNA or RNA are quite similar methods by the nature of their execution, albeit with subtle differences. The technique has a high phylogenetic resolution, and has provided many new insights to this day concerning microbial biodegradation of specific compounds and putative interrelationships of microbial activities with biogeochemical processes. This chapter provides an overview on the methodology, its caveats, and gives examples of applications for exploring the diversity of microbial hydrocarbon degraders in seep and other benthic habitats.

T. Gutierrez (✉)
Heriot-Watt University, Edinburgh, UK
e-mail: tony.gutierrez@hw.ac.uk

S. Kleindienst
Eberhard Karls University of Tübingen, Tübingen, Germany
e-mail: sara.kleindienst@uni-tuebingen.de

A. Teske and V. Carvalho (eds.), *Marine Hydrocarbon Seeps*,
Springer Oceanography, https://doi.org/10.1007/978-3-030-34827-4_10

Introduction

Our knowledge on the diversity of hydrocarbon-degrading bacteria in the ocean is far from complete, but recent advances in molecular biology techniques have marked a turning point towards achieving a better understanding of this. The development of next generation sequencing, which combines improved phylogenetic resolution with depth of coverage, has revolutionized our perception of the microbiome for a multitude of different environments on Earth (Gilbert and Dupont 2011; Jansson et al. 2012). However, in order to uncover new taxa of hydrocarbon-degraders in the field we require methods that couple phylogeny with the ability to assign, with a high level of confidence, whether the identified taxa are capable of degrading hydrocarbons. In fact, this correlation of microbial diversity and ecosystem function remains a major challenge for improving our understanding on the complexity of an ecosystem, especially in the ocean, where local and global biogeochemical processes affect and control microbial activities and vice versa (Glöckner et al. 2012). Notably, interest in research on marine oil spills and hydrocarbon degradation processes has intensified over the years and has, at present, reached unprecedented heights (Murphy et al. 2016).

A chapter on techniques to identify hydrocarbon-degraders in seep and other deep-sea habitats would be amiss without introducing the present state of knowledge on the diversity of these organisms (for details on anaerobic sulfate-reducing hydrocarbon-degraders see Chaper "Anaerobic Hydrocarbon-Degrading Sulfate-Reducing Bacteria at Marine Gas and Oil Seeps"). To begin, the diversity of aerobic hydrocarbon-degrading microorganisms in the global ocean is largely confined to the domain *Bacteria*. They are ubiquitous (Head et al. 2006; Yakimov et al. 2007), and mediate the degradation of much of the oil-hydrocarbons that enter the ocean (Leahy and Colwell 1990; Prince 2010; Atlas and Hazen 2011). As far as we know the marine environment is the only place on Earth where certain types of bacteria that specialize in the degradation of hydrocarbons, commonly referred to as obligate hydrocarbonoclastic bacteria (OHCB), are found. These organisms are distinguished by their ability to utilize hydrocarbons almost exclusively as a sole source of carbon and energy, and, with the exception of *Planomicrobium alkanoclasticum* which belongs to the phylum *Firmicutes*, all known genera of aerobic OHCB are *Gammaproteobacteria* (Gutierrez 2018). Known anaerobic hydrocarbon-degrading microorganisms in the marine environment mostly affiliate with sulfate-reducing *Deltaproteobacteria* or *Firmicutes* (Kleindienst et al. 2012, 2014; Orcutt et al. 2010; Widdel et al. 2010, Chapter "Anaerobic Hydrocarbon-Degrading Sulfate-Reducing Bacteria at Marine Gas and Oil Seeps"). On the other hand, the diversity of bacteria that can utilise hydrocarbons and non-hydrocarbon substrates as a source of carbon and energy—commonly referred to as generalist hydrocarbon-degraders—is orders of magnitude greater and spans across every major class of bacteria (Prince et al. 2010). Whilst the OHCB are highly specialised for utilising hydrocarbons, generalists are more metabolically versatile and capable of utilising a much wider spectrum of organic substrates. Often, however, the identification of hydrocarbon-degrading generalists is missed because evidence to substantiate their direct involvement in the degradation process is lacking. For example, the strong enrichment and dominant representation of a particular taxonomic group in a sequencing survey from an oil-contaminated environmental site is insufficient evidence to substantiate its direct involvement in the initial degradation and metabolism of hydrocarbons. Furthermore, cultivation-dependent techniques have likewise proven to be limited in their ability to reveal the near complete diversity of a microbial community in an environmental setting (Staley and Konopka 1985).

Stable isotope probing (SIP) is a sophisticated molecular technique for characterizing the roles of microorganisms in their natural environment without the necessity to isolate them in pure culture. SIP can be categorized as a semi-cultivation independent technique as it can be applied to enriched cultures in the laboratory, but also in the field. Pioneered by Boschker et al. (1998), it was first used to show the importance of type I methanotrophs in methane mineralization in lake sediments by analyzing the incorporation of ^{13}C from $^{13}CH_4$ into phospholipid fatty acids (PLFAs). SIP has subsequently been used for analyzing isotopically-labeled DNA (Radajewski et al. 2000), RNA (Manefield et al. 2002), and proteins (Jehmlich et al. 2008). The rationale to SIP is based on enriching a microbial community with an isotopically-labeled substrate—typically ^{13}C or ^{15}N, though more recently ^{18}O has also been used successfully (Schwartz 2007). During the so-called pulse incubation, members of the community that actively metabolize the labeled substrate will incorporate it into their biomolecules. The biomolecule(s) of interest (DNA, RNA, protein, or PLFAs) are then subsequently extracted for downstream phylogenetic analysis to identify the members of the community that incorporated the isotopic label. These "incorporators" are, hence, those that metabolized the labeled substrate. Compared to PLFA- and protein-based SIP, DNA- and RNA-based SIP techniques are most frequently applied in environmental microbiology studies—see Gutierrez-Zamora and Manefield (2010) for a review. DNA- and RNA-SIP offer a higher degree of phylogenetic resolution and functional information from the labeled nucleic acids (Neufeld et al. 2007a, 2008; Wischer et al. 2014), and will be the focus of this chapter for microbial hydrocarbon degradation studies in ocean systems.

DNA- and RNA-SIP have been used to investigate hydrocarbon-degrading microorganisms in marine environments, but considering the power of this technique it has scarcely been used for this purpose and within this environmental setting, and even less so in deep-sea environments. SIP is relatively costly, laborious, and technically sophisticated, and the challenges of working in the deep sea, not to mention the costs associated with research cruises that are prerequisite to acquiring the samples in the first place, gives SIP a less palatable selection in the repertoire of methods of a marine scientist. For those up to the challenge, the dawning of next generation sequencing approaches has unearthed an era where the diversity of hydrocarbon-degrading microbial populations and their genetic potential can be examined at an unprecedented scale, allowing for novel discoveries on this front. The close coupling of SIP with cutting-edge omics approaches (metagenomics, metatranscriptomics, metaproteomics, metabolomics), e.g. via

Illumina MiSeq/HiSeq sequencing, offers a powerful way to uncover more of the hidden diversity of hydrocarbon-degrading populations in the ocean, including their metabolic potential, and the chemical and physiological cues that control their activities.

SIP Methodology

DNA-SIP

A typical DNA-SIP experiment involves incubating an environmental sample (e.g. soil, water, sediment) with a ^{13}C-labeled compound that will act as the substrate for assimilation by the microbial population being targeted for identification in the sample. For example, to identify microorganisms capable of metabolizing *n*-hexadecane at a seep site, a sample of surficial sediment is pulsed with ^{13}C-labeled *n*-hexadecane. Ideally, uniformly-labeled ^{13}C-*n*-hexadecane is preferred over the partially-labeled compound as the former allows more label to be incorporated into biomass, and in shorter time scales. Longer incubation times are undesirable because it can allow cross-feeders—i.e. microbial members of the community that feed off the intermediates from the metabolism of the parent compound, or, off dead biomass from the primary metabolizers—to also incorporate the label (Fig. 1), thereby complicating the discrimination of member(s) of a community that were directly involved in the assimilation of the *n*-hexadecane. This time issue is one limitation of DNA-SIP since label incorporation into DNA requires cell replication. Therefore, it becomes necessary to use a tractable platform to monitor the

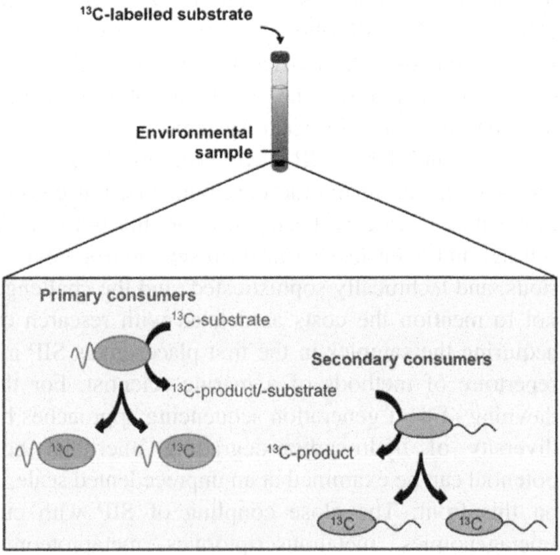

Fig. 1 A cross feeding network with primary substrate consumers and potential secondary metabolite consumers within a SIP experiment. Labeling of secondary metabolite consumers is a disadvantage of DNA- and RNA-SIP, which can be turned into an advantage by tracking the carbon flow, i.e. sampling several time points that cover the identification of primary consumers and potential secondary metabolite consumers

metabolism of the labeled substrate. Taking samples for SIP during the pulse incubation (i.e. sampling a time series) can help resolve the emergence of cross-feeders, whilst also providing insight into trophic networks of the microbial population by following the incorporation of the isotopic label into other members of the community that are not directly assimilating the labeled parent compound.

A SIP experiment employing a relatively easy-to-metabolize hydrocarbon substrate, such as naphthalene, may only require 24 h for aerobic naphthalene-degraders of a community to incorporate sufficient levels of the ^{13}C label into their DNA. However, for more complex, higher-molecular-weight hydrocarbon substrates, such as benzo[a]pyrene, or, for anaerobic hydrocarbon degraders, the pulse period will need to be (much) longer in order to obtain sufficient incorporation of the label into microbial biomass. For this reason, monitoring the disappearance of the substrate (due to its biodegradation/metabolism), the production of metabolic products (e.g. $^{13}CO_2$), or the incorporation of the stable isotope into biomass (e.g. ^{13}C-total organic carbon), is desirable, and provides a guidance tool for when to terminate the pulse incubation. In addition, monitoring the mineralization of the substrate, e.g. measuring $^{14}CO_2$ produced in incubations that are run in parallel with a radiolabeled (^{14}C) counterpart of the hydrocarbon substrate, can be used as guide for when most or all of the hydrocarbon has been mineralized. This acts as a proxy for when the isotopic label is near-complete or been fully incorporated into biomass.

Another consideration when performing SIP is deciding how much, or what concentration of labeled substrate should be used during the initial pulse. One of the hallmarks of SIP is that it averts the need to cultivate microorganisms in order to characterize their metabolic potential. In effect, the pulse can be carried out under conditions that are identical, or close, to those found in the field. It is, hence, advised to add the label at a concentration that reflects that found in the environment being studied. However, enrichment with in situ concentrations of a labeled substrate may be too low for DNA-SIP, in which case a highly (ideally uniformly) labeled substrate should be used, and at higher-than-in situ concentrations, in order to obtain sufficient labeling in a relatively short time scale to minimize complications from cross-feeding. On the other hand, using higher-than-in situ concentrations of substrate precludes the goal to replicate in situ conditions, since increasing the concentration of a hydrocarbon could potentially suppress the activities of some microbial members whilst enhance that of others. At the end of the day, the overarching aim of SIP is to identify the population in a community that are capable of metabolizing the labeled substrate, so a balance must be found for performing the pulse for a DNA-SIP experiment that takes into account substrate labeling, substrate concentration, and incubation time so that sufficient labeling of DNA is achieved for adequate separation.

At the termination of the pulse incubation, the DNA extracted of the whole microbial community comprises a mix of heavy (^{13}C-enriched) DNA from members of the community that metabolized the *n*-hexadecane, and of unlabeled light (^{12}C) DNA from the other members of the community. The success of DNA-SIP to identify microorganisms that assimilate an isotopically-labeled substrate is based on exploiting the increased buoyant density of the isotopically-labeled heavy DNA (belonging to the microbial population that incorporated the isotopic label) relative to unlabeled light DNA. The subtle difference in buoyant density between ^{13}C-labeled and unlabeled DNA (ca. 0.04 g/ml) is significant enough to achieve separation (Lueders et al. 2004), which is performed by isopycnic ultracentrifugation in a density gradient of cesium chloride. DNA of different densities migrates in the gradient and becomes localized where the density of the gradient equals its own. Uniformly ^{13}C-labeled DNA will increase in buoyant density by 0.036 g/ml (Birnie and Rickwood 1978). The use of a nucleic acid stain (e.g. ethidium bromide, or, SYBR-Safe), which is added to the cesium chloride-DNA mix prior to ultracentrifugation, provides a visual means to observe the position of the DNA banding pattern—the ^{13}C-enriched DNA band will be found further down in the gradient, whereas the ^{12}C DNA band will lie somewhere above it (Fig. 2). Whilst in theory only two DNA bands should be present within the gradient, in praxis the DNA is typically spread between these bands, sometimes even further above and below this range. Essentially, and most importantly for the researcher performing the experiment, the increasingly labeled DNA is found further down in the gradient. It needs to be accepted that it is practically unlikely that complete separation of heavy and light DNA can be achieved by isopycnic ultracentrifugation of DNA from SIP. Adding to this, the position of DNA in an isopycnic gradient is further

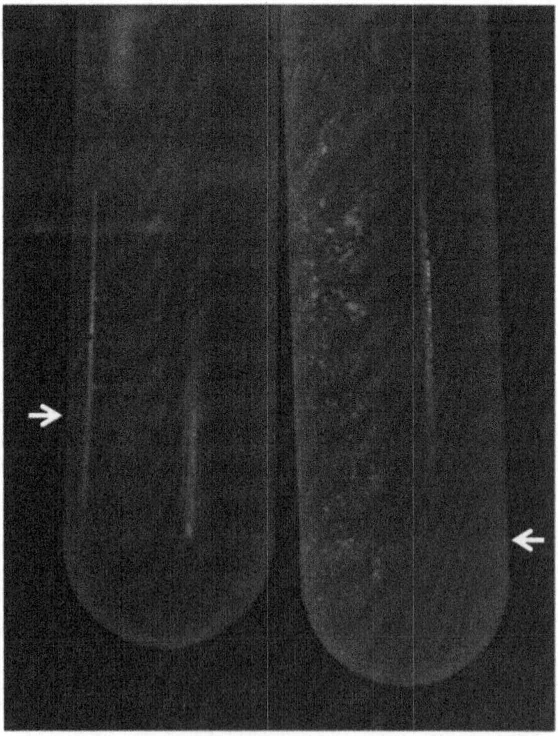

Fig. 2 Separation of light (tube on left) and heavy (tube on right) DNA by isopycnic ultracentrifugation in density gradients of cesium chloride from biomass recovered from a DNA-SIP experiment, using seawater enriched with uniformly ^{13}C-labelled *n*-hexadecane. The DNA in the gradients (arrows) was visualized using SYBR-Safe

complicated by variations in the G + C content of different members of a microbial community—the buoyant density of unlabeled DNA alone, due to G + C content variability, can be as much as 0.05 g/ml (Buckley et al. 2007; Youngblut and Buckley 2014).

The recovery of heavy and light DNA from the gradient can be achieved using a syringe, although fractionation of the gradient, whilst more laborious, offers more complete recovery of the DNA. Following this, the DNA is precipitated and washed to remove the cesium chloride, the nucleic acids stain, and any other impurities prior to using fingerprinting methods to analyze the 16S rRNA gene profiles of the DNA recovered. Quantitative PCR is used to quantify the distribution of total DNA in the density gradients. Comparing (^{12}C) control incubations with (^{13}C) SIP experiments, distinct peaks might point towards the fraction containing heavy DNA. The 16S rRNA gene fingerprinting analysis is a relatively quick and important step to unambiguously identify the fractions that comprise the heavy DNA band in the cesium chloride gradient relative to an unlabeled (^{12}C) control incubation. The most common fingerprinting methods are DGGE (denaturing gradient gel electrophoresis) and T-RFLP (terminal restriction fragment length polymorphism), and when coupled with clone library construction or a demultiplexing sequencing approach, such as barcoded-amplicon MiSeq or pyrosequencing, provide a powerful tool to uncovering the diversity and function of the microbial community (Neufeld et al. 2007b).

The extremely high phylogenetic resolution that is offered by DNA-SIP, and the relative ease in which DNA can be extracted from environmental samples (compared to RNA and protein extraction) make this technique, at present, the most widely used of SIP methodologies. To conduct a DNA-SIP experiment at sea in situ, such as at a seep site or similar benthic habitat, the requirement to know when to end the pulse, which is guided by monitoring the consumption of the isotopically-labeled substrate using an appropriate analysis for the respective substrate (e.g. GC/GC-MS, HPLC), makes this prospect somewhat daunting. A sufficient amount of the label (>15–20 atom%) needs to be incorporated in order to subsequently achieve separation of the ^{13}C-labeled and unlabeled DNA (Radajewski et al. 2000; Whiteley et al. 2007). Therefore, SIP performed in situ is possible, given it is accessible for the routine collection of samples to track consumption of the substrate on an hourly to daily basis. For example, Jeon et al. (2003) conducted a field-based SIP experiment with ^{13}C-naphthalene at a shallow unconfined aquifer near the western edge of the Hudson River, NY, revealing a novel bacterium with a distinctive dioxygenase; the study site was easily accessible for sample collection to monitor naphthalene consumption over a pulse period of only 8 h using GC/MS. This, however, would be impractical or near to impossible when the pulse incubation is running at several hundred to thousand meters below the sea surface. A sophisticated autonomous setup that is equipped with a substrate-monitoring instrument, and that is deployable and capable of operating autonomously on the sea floor, would provide an opportunity, though likely quite expensive, to instigate microbial hydrocarbon DNA-SIP studies in situ. DNA-SIP has been scarcely exploited for microbial hydrocarbon degradation studies at seep sites, or, at any deep-sea habitats in general. Consequently, for the named logistical and economic reasons, current studies are performed on samples from sediment cores or benthic grabs, which are recovered aboard a ship and slurries are produced for laboratory-based SIP experiments (Gutierrez et al. 2015).

RNA-SIP

RNA-SIP experiments are performed analogous to DNA-SIP experiments. The choice of sampling material and labeled substrate(s) as well as the SIP experimental setup is the same as for DNA-SIP. This includes ^{12}C-control experiments and analyses of substrate, product, and biomass of parallel incubations in order to get a proxy for labeled substrate turnover or incorporation (see section "DNA-SIP"). However, the incubation time of RNA-SIP experiments is shorter compared to DNA-SIP because RNA-SIP is much more sensitive, and cells do not need to divide for label incorporation. RNA-SIP focuses on active microbial populations from the complex microbial community due to the instability of RNA in environmental samples (Deutscher 2006) and, therefore, emphasizes metabolically-active populations even more. The rate of incorporation of the label into RNA is faster than into DNA (Manefield et al. 2002), and label incorporation of RNA-SIP requires a minimum of 15 atom% of ^{13}C (Molin and Givskov 1999) in order to achieve a separation of heavy (i.e., microbial populations that actively incorporated the isotope label) and light (i.e., microbial background populations) RNA.

Besides this, even more pronounced differences of RNA-SIP compared to DNA-SIP lie in the downstream protocol after the incubation has been terminated. All buffers and solutions are prepared for RNA work (e.g., using DEPC-treated solutions) and special care needs to be taken when working with RNA (RNase free working area). Because RNA is less stable compared to DNA, samples should be stored at −80 °C. Upon termination of the pulse, RNA is extracted with a protocol yielding high amounts of good quality RNA. Separation of heavy and light RNA is achieved by isopycnic ultracentrifugation in a density gradient of cesium trifluoracetate (Whiteley et al. 2007). Like for DNA, the position of RNA in an isopycnic gradient is also influenced by variations in the G + C content and secondary structure, so that the buoyant density of unlabeled RNA can be as much as 0.08 g/ml (Buckley et al. 2007).

Following the fractionation and recovery, the RNA fractions are likewise precipitated from the gradient matrix and washed. Subsequent downstream analyses are now again similar to DNA-SIP with the extra step of needing to reverse transcribing the RNA for quantitative PCR, for instance, to quantify the distribution of total RNA in the density gradients (Lueders et al. 2004). Typically, the same fingerprinting techniques (e.g. T-RFLP, DGGE) are then performed in combination with clone library construction or next generation sequencing, with protocols modified for RNA analyses, i.e. including again a reverse transcription step.

Coupling DNA-/RNA-SIP with Other Techniques

The ability to recover nucleic acids (DNA, RNA) that are enriched with an isotopic label from the assimilation of a ^{13}C-labeled hydrocarbon substrate offers unfathomed potential to uncover the near complete diversity of the microbial population in a complex community that is capable of metabolizing the substrate of interest, as well as to explore the genetic potential of their genomes. High-throughput sequencing approaches (e.g., barcoded-amplicon pyrosequencing and Illumina MiSeq sequencing) that are based on the 16S rRNA gene marker have revolutionized sequencing, providing a resolution of sequencing depth that is unprecedented since the dawning of this technology about a decade and a half ago. However, whilst sequencing surveys provide valuable information on 'who is there', the presence of any taxon identified does not necessarily allow to infer that these organisms are alive or active. Their putative activity in, for example, metabolizing a particular hydrocarbon substrate remains speculative when solely based on their relative abundance and phylogenetic affiliation to established hydrocarbon-degrading representative strains reported in the literature. Combining SIP with high-throughput sequencing of isotopically-enriched DNA or RNA, however, provides the possibility to receive an almost complete picture of the diversity of microorganisms in a complex community, for up to thousands of taxa, that incorporated an isotopic label of interest, i.e. proof of function. Methodological information on this SIP-metagenomics approach is available in a detailed protocol by Grob et al. (2015). The costs involved for this should not be shied upon, but when the funding is available the information gained can belittle the price paid. For example, high-throughput metagenomics approaches allow for less abundant ('rare species') to be identified (from extracted DNA or RNA), for genomes to be reconstructed (from extracted DNA), and for gene-mining of functional genes (from extracted DNA or RNA).

In a recent study by Dombrowski et al. (2016), heavy DNA, from DNA-SIP experiments performed on surface oil slicks and deep plume water samples collected during the active phase of the Deepwater Horizon oil spill, was metagenomically sequenced for genome reconstruction. The source DNA was derived from SIP incubations that had been pulsed with ^{13}C-labeled naphthalene, phenanthrene, or *n*-hexadecane. The DNA-SIP approach allowed for the identification of those microbial responders that demonstrably assimilated aromatic and alkane hydrocarbons during the spill, including some taxa that have so far escaped cultivation. The combined DNA-SIP and genomics approach expanded these findings to reveal the pathways for alkane and aromatic hydrocarbon degradation encoded by the reconstructed ^{13}C-enriched genomes. This revealed, for the first time in studies on the Deepwater Horizon spill, the existence of persistent pathway gaps for aromatic hydrocarbon degradation in all reconstructed genomes, alluding to a coordinated metabolic intra-dependency between these organisms with these incomplete pathways in order to explain the abilities of the whole community in responding to this massive and unprecedented oil spill. Analysis of these isotopically-enriched reconstructed genomes also uncovered other features of their metabolic and physiological capabilities and constraints, and provided a near-complete genetic and taxon-specific overview on metabolic potential for many of the major hydrocarbon-degrading responders during the spill.

The application of SIP alone, or even better in combination with metagenomics, offers the opportunity to enriching what little we currently know about the diversity and function of microbial hydrocarbon degraders in the deep ocean, especially in hydrocarbon seeps where these organisms have existed and thrived for millennia. Coupling these approaches will help uncover novel hydrocarbon-degrading taxa, and resolve their functions at the population level based on their ability to assimilate hydrocarbon substrates. This would also be aided by complementing these methods with other sophisticated methods, such as Raman microspectroscopy, nanometer-scale secondary ion mass spectrometry (nanoSIMS), and fluorescence in situ hybridization (FISH) (Warnecke and Hugenholtz 2007). In addition to advancing our knowledge on the diversity, physiology, and ecology of microbial hydrocarbon degraders in seeps and similar environments, there is considerable potential in combining RNA-/DNA-SIP with metagenomics to exploit the biotechnological potential of these and other microorganisms. For instance, hydrocarbon-degrading bacteria are commonly associated with the synthesis of surface-active agents (i.e., bio-surfactants or bio-emulsifiers)—chemicals that are of immense importance and used in almost every sector of modern industry today. The extreme conditions found at seep environments suggest that resident hydrocarbon degraders may have significant promise for the discovery of novel types of surface-active agents, enzymes, and bio-active agents with biotechnological potential. A major drawback here, however, is that SIP permits the analysis of only few samples at a time, and is thus a relatively slow process for screening. Achieving high-throughput to accelerate the large-scale discovery pipeline for new surface-active agents, enzymes, and other bio-active molecules using SIP (and in combination with metagenomics) will be a challenge that is likely to be initiated in the future.

While most of the DNA- and RNA-based SIP studies target phylogenetically informative genes and transcripts (i.e., 16S rRNA) aiming for the taxonomic identification of microorganisms involved in the assimilation of a labeled compound, complementary analyses of functional genes and transcripts further allow insights into metabolic abilities of active populations. For example, aerobic methane-oxidizing bacteria (methanotrophs) were identified in sediments from Lake Stechlin, Germany, and quantitative PCR targeting *pmoA* genes and transcripts, encoding a subunit of the key enzyme methane monooxygenase, was applied to track label incorporation into these marker genes and transcripts (Dumont et al. 2011). In another SIP study, CO_2 fixation coupled to ammonia oxidation was examined in soil, and sequencing of archaeal acetyl-CoA/propionyl-CoA carboxylase (*accA*/*pccB*) transcripts from heavy fractions suggested that CO_2 fixation was performed using the 3-hydroxypropionate-4-hydroxybutyrate cycle (Pratscher et al. 2011). Furthermore, labeling-assisted targeted metatranscriptomes of aerobic methanotrophs were retrieved from lake sediment (Dumont et al. 2013), demonstrating opportunities to explore metatranscriptomes of the active target population from SIP experiments, which might be promising for future SIP approaches (Lueders et al. 2016). Whilst these studies were not specifically investigating hydrocarbon-degraders, the examples highlight sequence-based methods targeting functional genes and transcripts as techniques that could be complemented with DNA- and RNA-based SIP to uncover more information about natural microbial systems.

Applications to Benthic Systems

One of the first studies to apply DNA-SIP to a marine habitat, and herein coupled with metagenomics, employed ^{13}C-glycerol to identify members of the community in marine sediment that contained B12-dependent glycerol- and diol-dehydratases (Schwarz et al. 2006). In a follow-up study, Neufeld et al. (2008) exposed surface seawater to ^{13}C-methanol in DNA-SIP coupled with multiple displacement amplification of the isolated 'heavy' DNA to identify a *Methylophaga*-related methanol-dehydrogenase gene cluster. With respect to the application of SIP to study microbial hydrocarbon degraders, the majority of studies on this have focused on terrestrial environments, such as soils or bioreactors (for a review see Lueders 2010). Less than a handful of studies have used SIP to investigate these organisms in marine environments, and herein mainly in the water column (Dombrowski et al. 2016; Gutierrez et al. 2011, 2013; Mishamandani et al. 2014). As noted earlier, the seafloor is a challenging environment to study, and increasingly so the deeper we venture and the more demanding the experimental approach we want to apply. Performing SIP in situ in the deep is not inconceivable (Enge et al. 2011; Jeffreys et al. 2013; Stratmann et al. 2018; Sweetman and Witte 2008; Sweetman et al. 2019; Witte et al. 2003), but it is riddled with significant challenges that make it a rarely-performed experiment at depth. For now, the vast majority of SIP enthusiasts are left with performing SIP in the deep ex situ, by collecting sediment samples during research cruises and using them as slurries to carry out the pulse in a laboratory setting. A few examples on the application of DNA- and RNA-SIP focusing on aerobic and anaerobic hydrocarbon degraders in deep-sea habitats follows below.

Aerobic SIP

As already mentioned, the use of SIP with hydrocarbon substrates has been scarcely applied to ocean systems, and less even to deep-sea environments. DNA-SIP was used to investigate bacteria that can degrade polycyclic aromatic hydrocarbons (PAH) under aerobic conditions in sediments from Guaymas Basin with ^{13}C-phenanthrene (Gutierrez et al. 2015). Situated in the Gulf of California, the Guaymas Basin lies $\sim$2000 m below the sea surface; it is a deep-sea model system for studying the diversity of hydrocarbon-degrading organisms (Teske et al. 2014) where aromatic/naphthenic hydrocarbon levels, including PAHs, can constitute up to 30% of the oil (Bazylinski et al. 1988). From SIP experiments conducted on sediment core-derived slurries, the dominant organism identified in clone libraries constructed from ^{13}C-enriched bacterial DNA (from phenanthrene enrichments) was *Cycloclasticus* (Gutierrez et al. 2015). A representative *Cycloclasticus* isolate was successfully cultivated from the sediment slurries that shared 99.9% 16S rRNA gene sequence identity to the SIP-identified clone and was shown to degrade phenanthrene. In the same study, strains of *Halomonas*, *Thalassospira* and *Lutibacterium* were also isolated with demonstrable phenanthrene-degrading capacity from Guaymas Basin sediment. This study showed the value of coupling SIP with cultivation methods to identify and expand on the known diversity of PAH-degrading bacteria in the Guaymas Basin.

Anaerobic SIP

Anaerobic hydrocarbon degraders, i.e. butane and dodecane degraders, from marine seep sediments were identified under sulfate-reducing conditions using DNA-SIP and RNA-SIP (Kleindienst et al. 2014). Sediment samples were taken at sites where anaerobic hydrocarbon degraders were hypothesized to be active: seeps at the Amon mud volcano in the Mediterranean Sea, emitting short-chain alkanes (gaseous C_1–C_5 alkanes) and hydrocarbon seeps at Guaymas Basin in the Gulf of California (Chap. 3), characterized by complex hydrocarbons (gaseous alkanes, longer-chain alkanes, and aromatics) (Bazylinski et al. 1988). Sulfate-reducing rates were elevated at the Amon mud volcano and at Guaymas Basin seeps, indicating active sulfate-reducing activities and making these sites ideal to identify anaerobic hydrocarbon degraders from contrasting, i.e. gas and hydrocarbon, marine seep sediments.

Sediment cores were taken at the Amon mud volcano and Guaymas Basin and stored anoxically and cold (4 °C) prior to the initiation of SIP experiments. Since precious sediment cores are limited in sediment material and heterogeneous, pooling and carefully mixing the sediments under anoxic conditions (e.g. in an anoxic chamber) might be a preferred option to obtain enough sampling material that is subsequently divided into incubation vials for parallel setups. Pooling sediment cores is also reasonable since most SIP studies do not target spatial resolution of the identified active population in the sediment cores. However, the activity of the target population might be quite different among sediment cores and even more so among seep sites from the same habitat. For this reason, a short pre-incubation might be very useful to determine sediment cores with elevated microbial activities that are promising for SIP studies. Kleindienst et al. (2014) used such pre-incubations to identify sediment cores with highest alkane-degrading microbial activities based on sulfide and $^{13}CO_2$ production, the products of sulfate reduction and alkane oxidation. This was an important strategy to select promising sediment material for SIP experiments before adding expensive ^{13}C-labelled substrates.

For the experimental design of the SIP experiment, Kleindienst et al. (2014) then also selected three sampling points, including initial time points before cross feeding began, in order to be able to identify primary consumers and to track the carbon-flow through the microbial community. Since ideal time points for sampling were not predictable, Kleindienst et al. (2014) carefully monitored sulfide production in incubation vials with and without substrate amendment as a measure for the activity of hydrocarbon-degrading sulfate-reducing bacteria. Ultimately, the sampling points of the SIP incubation by Kleindienst et al. (2014) varied among seep sediments investigated (i.e. Amon mud volcano and Guaymas Basin), and substrates used (i.e. butane and dodecane), emphasizing that the experimental design needs to be adjusted to the activity of the microbial target population. Since this activity depends on the abundance of the target population, its metabolic pathway, and on environmental factors (e.g. temperature, nutrients, substrate and electron acceptor availability), sampling points may vary substantially among different experiments. Also, depending on whether DNA-SIP or RNA-SIP is anticipated, sampling might be performed after days, weeks, or even months. Another critical aspect of the SIP experimental design is the amount of labeled and unlabeled substrate to add, which is a trade-off between in situ conditions and the amount needed to achieve sufficient labeling for DNA- and RNA-SIP experiments (see Sects. 1 and 2 in this chapter). Therefore, SIP studies typically apply a higher substrate concentration compared to their concentrations found in situ. The choice of incubation temperature is another important factor for the experimental design. Kleindienst et al. (2014) incubated their samples slightly above in situ temperatures (Amon mud volcano incubation at 20 °C, compared to 14 °C in situ; Guaymas Basin incubation at 28 °C, compared to seep temperature gradients of 3–50 °C in situ) in order to achieve a stimulation of microbial activities. Finally, the addition of nutrients and excess electron acceptor(s) might be important to ensure that the experiment is not limited by these factors. All these choices are a trade-off between requirements of the SIP method and in situ-like incubation conditions.

At the Amon mud volcano and Guaymas Basin, four groups of alkane-degrading sulfate-reducing bacteria were identified using DNA-SIP and RNA-SIP (for details see Chapter "Anaerobic Hydrocarbon-Degrading Sulfate-Reducing Bacteria at Marine Gas and Oil Seeps"). These groups, named SCA1, SCA2, LCA1, and LCA2, were found affiliated with the *Desulfosarcina/Desulfococcus* clade of the *Deltaproteobacteria* (Kleindienst et al. 2014). Three of the four groups comprised uncultivated members, emphasizing one of the greatest advantages of nucleic-acid SIP, which is the identification of uncultivated members of complex microbial communities that are key players of a specific metabolic function such as hydrocarbon biodegradation.

Summary

The application of density-resolved DNA/RNA fractions with next-generation sequencing methodologies represents a major recent advance in DNA-/RNA-SIP to unravel specific metabolic capabilities, ecophysiologies, and interactions of microorganisms in natural systems. This coupling of SIP with 'omics' techniques has shown significant potential as it provides functional context to taxonomic identity that can be substantiated through label incorporation. When combined also with other contemporary cutting-edge technologies (e.g. Raman microspectroscopy, nanoSIMS, FISH etc.), the future of SIP is guaranteed with high productivity. The number of studies that have applied SIP to study hydrocarbon degradation processes in deep-sea environments has currently not yet reached double digits. This is largely due to a few main reasons, one of which is the costs involved and the sophistication required with performing SIP—this is not a technique that lends itself as easy to learn independently. Until today, only few research groups around the world have established SIP in their laboratories or use large-scale metagenomics in environmental studies. But the most important reason limiting the use of SIP as experimental standard for deep-sea research is the difficulty involved with setting up and running a ^{13}C-labeling pulse in situ at hundreds to thousands of meters below the sea surface. For now, to study hydrocarbon-degradation processes in deep-sea environments we will need to make do with performing SIP in a laboratory setting, using sediment (cores or slurries) collected during research cruises. Once this obstacle is overcome, such as by the development of remotely-operated instruments that can operate SIP in situ, we may look forward to a new area in SIP and the discoveries this will uncover.

References

Atlas RM, Hazen TC (2011) Oil biodegradation and bioremediation: a tale of the two worst spills in U.S. Environ Sci Technol 45:6709–6715

Bazylinski DA, Farrington JW, Jannasch HW (1988) Hydrocarbons in surface sediments from a Guaymas Basin hydrothermal vent site. Org Geochem 12:547–558

Birnie GD, Rickwood D (1978) Centrifugal separations in molecular and cell biology. Butterworths, Boston

Boschker HTS, Nold SC, Wellsbury P, Bos D, de Graaf W, Pel R, Parkes RJ, Cappenberg TE (1998) Direct linking of microbial populations to specific biogeochemical processes by ^{13}C-labelling of biomarkers. Nature 392:801–805

Buckley DH, Huangyutitham V, Hsu SF, Nelson TA (2007) Stable isotope probing with ^{15}N achieved by disentangling the effects of genome G + C content and isotope enrichment on DNA density. Appl Environ Microbiol 73:3189–3195

Deutscher MP (2006) Degradation of RNA in bacteria: comparison of mRNA and stable RNA. Nucleic Acids Res 34:659–666

Dombrowski N, Donaho JA, Gutierrez T, Seitz KW, Teske AP, Baker BJ (2016) Reconstructing metabolic pathways of hydrocarbon-degrading bacteria from the Deepwater Horizon oil spill. Nat Microbiol 1:16057

Dumont MG, Pommerenke B, Casper P, Conrad R (2011) DNA-, rRNA- and mRNA-based stable isotope probing of aerobic methanotrophs in lake sediment. Environ Microbiol 13:1153–1167

Dumont MG, Pommerenke B, Casper P (2013) Using stable isotope probing to obtain a targeted metatranscriptome of aerobic methanotrophs in lake sediment. Environ Microbiol Rep 5:757–764

Enge AJ, Nomaki H, Ogawa NO, Witte U, Moeseneder MM, Lavik G, Ohkouchi N, Kitazato H, Kučera M, Heinz P (2011) Response of the foraminiferal community to a simulated phytodetritus pulse in the abyssal North Pacific. Mar Ecol Prog Ser 438:129–142

Gilbert JA, Dupont CL (2011) Microbial Metagenomics: beyond the Genome. Ann Rev Mar Sci 3:347–371

Glöckner FO, Gasol JM, McDonough N, Calewaert J-B (2012) Marine microbial diversity and its role in ecosystem functioning and environmental change. In: Calewaert JB, McDonough N (eds) Marine board-ESF position paper 17. Marine Board-ESF, Ostend

Grob C, Taubert M, Howat AM, Burns OJ, Chen Y, Neufeld JD, Murrell CJ (2015) Generating enriched metagenomes from active microorganisms with DNA stable isotope probing. In: McGenity T, Timmis K, Nogales B (eds) Hydrocarbon and lipid microbiology protocols. Springer Protocols Handbooks, Springer, Berlin, Heidelberg

Gutierrez T (2018) Marine, aerobic hydrocarbon-degrading *Gammaproteobacteria*—overview. In: McGenity TJ, Prince R (eds) Handbook of hydrocarbon and lipid microbiology. Taxonomy, genomics and ecophysiology of hydrocarbon-degrading microbes. Springer International Publishing

Gutierrez T, Singleton DR, Aitken MD, Semple KT (2011) Stable-isotope probing of an algal bloom identifies uncultivated members of the *Rhodobacteraceae* associated with low molecular-weight PAH degradation. Appl Environ Microbiol 77:7856–7860

Gutierrez T, Singleton DR, Berry D, Yang T, Aitken MD, Teske A (2013) Hydrocarbon-degrading bacteria enriched by the *Deepwater Horizon* oil spill identified by cultivation and DNA-SIP. ISME J 7:2091–2104

Gutierrez T, Biddle JF, Teske A, Aitken MD (2015) Cultivation-dependent and cultivation-independent characterization of hydrocarbon-degrading bacteria in Guaymas Basin sediments. Front Microbiol 6:695

Gutierrez-Zamora M-L, Manefield M (2010) An appraisal of methods for linking environmental processes to specific microbial taxa. Rev Environ Sci Biotechnol 9:153–185

Head IM, Jones DM, Röling WFM (2006) Marine microorganisms make a meal of oil. Nat Rev Microbiol 4:173–182

Jansson JK, Neufeld JD, Moran MA, Gilbert JA (2012) Omics for understanding microbial functional dynamics. Environ Microbiol 14:1–3

Jeffreys RM, Burke C, Jamieson AJ, Narayanaswamy BE, Ruhl HA, Smith KLS, Witte U (2013) Feeding preferences of abyssal macrofauna inferred from in situ pulse chase experiments. PLoS ONE 8:e80510

Jehmlich N, Schmidt F, von Bergen M, Richnow HH, Vogt C (2008) Protein-based stable isotope probing (Protein-SIP) reveals active species within anoxic mixed cultures. ISME J 2:1122–1133

Jeon CO, Park W, Padmanabhan P, DeRito C, Snape JR, Madsen EL (2003) Discovery of a bacterium, with distinctive dioxygenase, that is responsible for in situ biodegradation in contaminated sediment. Proc Natl Acad Sci USA 100:13591–13596

Kleindienst S, Ramette A, Amann R, Knittel K (2012) Distribution and in situ abundance of sulfate-reducing bacteria in diverse marine hydrocarbon seep sediments. Environ Microbiol 14:2689–2710

Kleindienst S, Herbst F-A, Stagars M, von Netzer F, von Bergen M, Seifert J et al (2014) Diverse sulfate-reducing bacteria of the *Desulfosarcina/Desulfococcus* clade are the key alkane degraders at marine seeps. ISME J 8:2029–2044

Leahy JG, Colwell RR (1990) Microbial degradation of hydrocarbons in the environment. Microbiol Rev 54:305–315

Lueders T (2010) Stable isotope probing of hydrocarbon-degraders. In: Timmis KN (ed) Handbook of hydrocarbon and lipid microbiology. Springer, Berlin, pp 4011–4026

Lueders T, Manefield M, Friedrich MW (2004) Enhanced sensitivity of DNA- and rRNA-based stable isotope probing by fractionation and quantitative analysis of isopycnic centrifugation gradients. Environ Microbiol 6:73–78

Lueders T, Dumont MG, Bradford L, Manefield M (2016) RNA-stable isotope probing: from carbon flow within key microbiota to targeted transcriptomes. Curr Opin Biotechnol 41:83–89

Manefield M, Whiteley AS, Griffiths RI, Bailey MJ (2002) RNA stable isotope probing, a novel means of linking microbial community function to phylogeny. Appl Environ Microbiol 68:5367–5373

Mishamandani S, Gutierrez T, Aitken MD (2014) DNA-based stable isotope probing coupled with cultivation methods implicates *Methylophaga* in hydrocarbon degradation. Front Microbiol 5:76

Molin S, Givskov M (1999) Application of molecular tools for in situ monitoring of bacterial growth activity. Environ Microbiol 1:383–391

Murphy D, Gemmell B, Vaccari L, Li C, Bacosa H, Evans M, Gemmell C, Harvey T, Jalali M, Neipa THR (2016) An in-depth survey of the oil spill literature since 1968: long term trends and changes since *Deepwater Horizon*. Mar Pollut Bull 113:371–379

Neufeld JD, Schafer H, Cox MJ, Boden R, McDonald IR, Murrell JC (2007a) Stable-isotope probing implicates *Methylophaga* spp. and novel *Gammaproteobacteria* in marine methanol and methylamine metabolism. ISME J 1:480–491

Neufeld JD, Vohra J, Dumont MG, Lueders T, Manefield M, Friedrich MW, Murrell JC (2007b) DNA stable-isotope probing. Nat Protoc 2:860–866

Neufeld JD, Chen Y, Dumont MG, Murrell JC (2008) Marine methylotrophs revealed by stable-isotope probing, multiple displacement amplification and metagenomics. Environ Microbiol 10:1526–1535

Orcutt BN, Joye SB, Kleindienst S, Knittel K, Ramette A, Reitz A et al (2010) Impact of natural oil and higher hydrocarbons on microbial diversity, distribution, and activity in Gulf of Mexico cold-seep sediments. Deep-Sea Res II 57:2008–2021

Pratscher J, Dumont MG, Conrad R (2011) Ammonia oxidation coupled to CO_2 fixation by archaea and bacteria in an agricultural soil. Proc Natl Acad Sci USA 108:4170–4175

Prince RC (2010) Bioremediation of marine oil spills. In: McGenity T, Timmis KN (ed) Handbook of hydrocarbon and lipid microbiology. Taxonomy, genomics and ecophysiology of hydrocarbon-degrading microbes. Springer, Berlin

Prince RC, Grammain A, McGenity TJ (2010) The microbes. In: Timmis KN, McGenity T (eds) Handbook of hydrocarbon and lipid microbiology. Prokaryotic hydrocarbon degraders. Springer, Berlin

Radajewski S, Ineson P, Parekh NR, Murrell JC (2000) Stable-isotope probing as a tool in microbial ecology. Nature 403:646–649

Schwartz E (2007) Characterization of growing microorganisms in soil by stable-isotope probing with $H_2^{18}O$. Appl Environ Microbiol 73:2541–2546

Schwarz S, Waschkowitz T, Daniel R (2006) Enhancement of gene detection frequencies by combining DNA-based stable-isotope probing with the construction of metagenomic DNA libraries. World J Microbiol Biotechnol 22:363–368

Staley JT, Konopka A (1985) Measurement of in situ activities of nonphotosynthetic microorganisms in aquatic and terrestrial habitats. Ann Rev Microbiol 39:321–346

Stratmann T, Mevenkamp L, Sweetman AK, Vanreusel A, van Oevelen D (2018) Has phytodetritus processing by an abyssal soft-sediment community recovered 26 years after an experimental disturbance? Front Mar Sci 5:59

Sweetman AK, Witte U (2008) Response of an abyssal macrofaunal community to a phytodetrital pulse. Mar Ecol Prog Ser 355:73–84

Sweetman AK, Smith CR, Shulse CN, Maillot B, Lindh M, Church MJ, Meyer KS, van Oevelen D, Stratmann T, Gooday AJ (2019) Key role of bacteria in the short-term cycling of carbon at the abyssal seafloor in a low particulate organic carbon flux region of the eastern Pacific Ocean. Limnol Oceanogr 64:694–713

Teske A, Callaghan AV, LaRowe DE (2014) Biosphere frontiers of subsurface life in the sedimented hydrothermal system of Guaymas Basin. Frontiers Microbiol 5:362

Warnecke F, Hugenholtz P (2007) Building on basic metagenomics with complementary technologies. Genome Biol 8:231

Whiteley AS, Thomson B, Lueders T, Manefield M (2007) RNA stable-isotope probing. Nat Protoc 2:838–844

Widdel F, Knittel K, Galushko A (2010) Anaerobic hydrocarbon-degrading microorganisms: an overview. In: Timmis KN, McGenity T, van der Meer TJR, de Lorenzo V (eds) Handbook of hydrocarbon and lipid microbiology. Springer, Berlin

Wischer D, Kumaresan D, Johnston A, El Khawand M, Stephenson J, Hillebrand-Voiculescu AM, Chen Y, Murrell CJ (2014) Bacterial metabolism of methylated amines and identification of novel methylotrophs in Movile Cave. ISME J 9:195–206

Witte U, Wenzhöfer F, Sommer S, Boetius A, Heinz P, Aberle N, Sand M, Cremer A, Abraham W-R, Jørgensen BB, Pfannkuche O (2003) In situ experimental evidence of the fate of a phytodetritus pulse at the abyssal sea floor. Nature 424:763–766

Yakimov MM, Timmis KN, Golyshin PN (2007) Obligate oil-degrading marine bacteria. Curr Opin Biotechnol 18:257–266

Youngblut ND, Buckley DH (2014) Intra-genomic variation in G + C content and its implications for DNA stable isotope probing. Environ Microbiol Rep 6:767–775